T0350600

FUNDAMENTALS
OF STOCHASTIC
NETWORKS

FUNDAMENTALS OF STOCHASTIC NETWORKS

OLIVER C. IBE
University of Massachusetts
Lowell, Massachusetts

A JOHN WILEY & SONS, INC., PUBLICATION

Published by John Wiley & Sons, Inc., Hoboken, New Jersey.
Published simultaneously in Canada.

For general information on our other products and services or for technical support, please
contact our Customer Care Department within the United States at (800) 762-2974, outside the
United States at (317) 572-3993 or fax (317) 572-4002.

Wiley also publishes its books in a variety of electronic formats. Some content that appears in
print may not be available in electronic formats. For more information about Wiley products,
visit our web site at www.wiley.com.

Library of Congress Cataloging-in-Publication Data:

Ibe, Oliver C. (Oliver Chukwudi), 1947–
  Fundamentals of stochastic networks / Oliver C. Ibe.
    p. cm.
  Includes bibliographical references and index.
  ISBN 978-1-118-06567-9 (cloth)
  1. Queuing theory. 2. Stochastic analysis. I. Title.
  QA274.8.I24 2011
  519.2'2–dc22
                                              2011007713

Printed in the United States of America

oBook ISBN: 9781118092972
ePDF ISBN: 9781118092996
ePub ISBN: 9781118092989

10 9 8 7 6 5 4 3 2 1

CONTENTS

PREFACE

This book brings into one volume two network models that can be broadly classified as *queueing network models* and *graphical network models*. Queueing networks are systems where customers move among service stations where they receive service. Usually, the service times and the order in which customers visit the service stations are random. The order in which service is received at the service stations is governed by a probabilistic routing schedule. Queueing networks are popularly used in traffic modeling in computer and telecommunications networks, transportation systems, and manufacturing networks. Graphical models are systems that use graphs to model different types of problems. They include Bayesian networks, which are also called *directed graphical models*, Boolean networks, and random networks. Graphical models are used in statistics, data mining, and social networks.

The need for a book of this nature arises from the fact that we live in an era of interdisciplinary studies and research activities when both networks are becoming important in areas that they were not originally used. Thus, any person involved in such interdisciplinary studies or research activities needs to have a good understanding of both types of networks. This book is intended to meet this need.

The book is organized into three parts. The first part, Chapters 1 and 2, deals with the basic concepts of probability (Chapter 1) and stochastic processes (Chapter 2). The second part, Chapters 3–6, deals with queueing systems. Specifically, Chapter 3 deals with basic queueing theory, particularly a class of queueing systems that we refer to as Markovian queueing systems. Chapter 4 deals with advanced queueing systems, particularly the non-Markovian queueing systems. Chapter 5 deals with queueing networks, and Chapter 6 deals with approximations of queueing networks. The third part, Chapters 7–10, deals

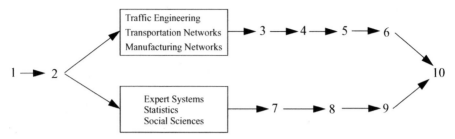

Figure 1 Precedence relations of chapters.

with graphical models. Chapter 7 deals with an introduction to graph theory, Chapter 8 deals with Bayesian networks, Chapter 9 deals with Boolean networks, and Chapter 10 deals with random networks.

The book is self-contained and is written with a view to circumventing the proof–theorem format that is traditionally used in stochastic systems modeling books. It is intended to be an introductory graduate text on stochastic networks and presents the basic results without much emphasis on proving theorems. Thus, it is designed for science and engineering applications. Students who have an interest in traffic engineering, transportation, and manufacturing networks will need to cover parts 1 and 2 as well as Chapter 10 in part 3, while students with an interest in expert systems, statistics, and social sciences will need to cover parts 1 and 3. The precedence relations among the chapters are shown in Figure 1.

ACKNOWLEDGMENTS

I would like to express my sincere gratitude to my wife, Christie, for bearing with my writing yet another book. She has been my greatest fan when it comes to my writing books. I would also like to acknowledge the encouraging words from our children Chidinma, Ogechi, Amanze, and Ugonna. I would like to express my sincere gratitude to my editor, Susanne Steitz-Filler, for her encouragement and for checking on me regularly to make sure that we met the deadlines. Finally, I would like to thank the anonymous reviewers for their useful comments and suggestions that helped to improve the quality of the book.

OLIVER C. IBE

Lowell, Massachusetts
January 2011

1

BASIC CONCEPTS IN PROBABILITY

1.1 INTRODUCTION

The concepts of *experiments* and *events* are very important in the study of probability. In probability, an experiment is any process of trial and observation. An experiment whose outcome is uncertain before it is performed is called a *random* experiment. When we perform a random experiment, the collection of possible elementary outcomes is called the *sample space* of the experiment, which is usually denoted by Ω. We define these outcomes as elementary outcomes because exactly one of the outcomes occurs when the experiment is performed. The elementary outcomes of an experiment are called the *sample points* of the sample space and are denoted by w_i, $i = 1$, $2, \ldots$ If there are n possible outcomes of an experiment, then the sample space is $\Omega = \{w_1, w_2, \ldots, w_n\}$. An *event* is the occurrence of either a prescribed outcome or any one of a number of possible outcomes of an experiment. Thus, an event is a subset of the sample space.

1.2 RANDOM VARIABLES

Consider a random experiment with sample space Ω. Let w be a sample point in Ω. We are interested in assigning a real number to each $w \in \Omega$. A random variable, $X(w)$, is a single-valued real function that assigns a real number,

Fundamentals of Stochastic Networks, First Edition. Oliver C. Ibe.
© 2011 John Wiley & Sons, Inc. Published 2011 by John Wiley & Sons, Inc.

called the value of $X(w)$, to each sample point $w \in \Omega$. That is, it is a mapping of the sample space onto the real line.

Generally a random variable is represented by a single letter X instead of the function $X(w)$. Therefore, in the remainder of the book we use X to denote a random variable. The sample space Ω is called the *domain* of the random variable X. Also, the collection of all numbers that are values of X is called the *range* of the random variable X.

Let X be a random variable and x a fixed real value. Let the event A_x define the subset of Ω that consists of all real sample points to which the random variable X assigns the number x.

That is,

$$A_x = \{w | X(w) = x\} = [X = x].$$

Since A_x is an event, it will have a probability, which we define as follows:

$$p = P[A_x].$$

We can define other types of events in terms of a random variable. For fixed numbers x, a, and b, we can define the following:

$$[X \le x] = \{w | X(w) \le x\},$$
$$[X > x] = \{w | X(w) > x\},$$
$$[a < X < b] = \{w | a < X(w) < b\}.$$

These events have probabilities that are denoted by

- $P[X \le x]$ is the probability that X takes a value less than or equal to x.
- $P[X > x]$ is the probability that X takes a value greater than x; this is equal to $1 - P[X \le x]$.
- $P[a < X < b]$ is the probability that X takes a value that strictly lies between a and b.

1.2.1 Distribution Functions

Let X be a random variable and x be a number. As stated earlier, we can define the event $[X \le x] = \{x | X(w) \le x\}$. The distribution function (or the cumulative distribution function [CDF]) of X is defined by:

$$F_X(x) = P[X \le x] \quad -\infty < x < \infty.$$

That is, $F_X(x)$ denotes the probability that the random variable X takes on a value that is less than or equal to x. Some properties of $F_X(x)$ include:

1. $F_X(x)$ is a nondecreasing function, which means that if $x_1 < x_2$, then $F_X(x_1) \le F_X(x_2)$. Thus, $F_X(x)$ can increase or stay level, but it cannot go down.

2. $0 \le F_X(x) \le 1$
3. $F_X(\infty) = 1$
4. $F_X(-\infty) = 0$
5. $P[a < X \le b] = F_X(b) - F_X(a)$
6. $P[X > a] = 1 - P[X \le a] = 1 - F_X(a)$

1.2.2 Discrete Random Variables

A discrete random variable is a random variable that can take on at most a countable number of possible values. For a discrete random variable X, the *probability mass function* (PMF), $p_X(x)$, is defined as follows:

$$p_X(x) = P[X = x].$$

The PMF is nonzero for at most a countable or countably infinite number of values of x. In particular, if we assume that X can only assume one of the values $x_1, x_2, \ldots, x_n$, then:

$$p_X(x_i) \ge 0 \quad i = 1, 2 \ldots, n,$$
$$p_X(x) = 0 \quad \text{otherwise.}$$

The CDF of X can be expressed in terms of $p_X(x)$ as follows:

$$F_x(x) = \sum_{k \le x} p_X(k).$$

The CDF of a discrete random variable is a step function. That is, if X takes on values $x_1, x_2, x_3, \ldots$, where $x_1 < x_2 < x_3 < \ldots$, then the value of $F_X(x)$ is constant in the interval between x_{i-1} and x_i and then takes a jump of size $p_X(x_i)$ at x_i, $i = 2, 3, \ldots$. Thus, in this case, $F_X(x)$ represents the sum of all the probability masses we have encountered as we move from $-\infty$ to x.

1.2.3 Continuous Random Variables

Discrete random variables have a set of possible values that are either finite or countably infinite. However, there exists another group of random variables that can assume an uncountable set of possible values. Such random variables are called continuous random variables. Thus, we define a random variable X to be a continuous random variable if there exists a nonnegative function $f_X(x)$, defined for all real $x \in (-\infty, \infty)$, having the property that for any set A of real numbers,

$$P[X \in A] = \int_A f_X(x) dx.$$

The function $f_X(x)$ is called the *probability density function* (PDF) of the random variable X and is defined by:

$$f_X(x) = \frac{dF_X(x)}{dx}.$$

The properties of $f_X(x)$ are as follows:

1. $f_X(x) \geq 0$
2. Since X must assume some value, $\int_{-\infty}^{\infty} f_X(x)dx = 1$
3. $P[a \leq X \leq b] = \int_a^b f_X(x)dx$, which means that $P[X = a] = \int_a^a f_X(x)dx = 0$. Thus, the probability that a continuous random variable will assume any fixed value is zero.
4. $P[X < a] = P[X \leq a] = F_X(a) = \int_{-\infty}^a f_X(x)dx$

1.2.4 Expectations

If X is a random variable, then the *expectation* (or *expected value* or *mean*) of X, denoted by $E[X]$, is defined by:

$$E[X] = \begin{cases} \sum_i x_i p_X(x_i) & X \quad \text{discrete} \\ \int_{-\infty}^{\infty} x f_X(x)dx & X \quad \text{continuous} \end{cases}$$

Thus, the expected value of X is a weighted average of the possible values that X can take, where each value is weighted by the probability that X takes that value. The expected value of X is sometimes denoted by $\bar{X}$.

1.2.5 Moments of Random Variables and the Variance

The nth moment of the random variable X, denoted by $E[X^n] = \overline{X^n}$, is defined by:

$$E[X^n] = \overline{X^n} = \begin{cases} \sum_i x_i^n p_X(x_i) & X \quad \text{discrete} \\ \int_{-\infty}^{\infty} x^n f_x(x)dx & X \quad \text{continuous} \end{cases}$$

for $n = 1, 2, 3, \ldots$. The first moment, $E[X]$, is the expected value of X.

We can also define the *central moments* (or *moments about the mean*) of a random variable. These are the moments of the difference between a random variable and its expected value. The nth central moment is defined by

$$E\left[(X - \bar{X})^n\right] = \overline{(X - \bar{X})^n} = \begin{cases} \sum_i (x_i - \bar{X})^n p_X(x_i) & X \quad \text{discrete} \\ \int_{-\infty}^{\infty} (x - \bar{X})^n f_X(x)dx & X \quad \text{continuous} \end{cases}$$

The central moment for the case of $n = 2$ is very important and carries a special name, the *variance*, which is usually denoted by σ_X^2. Thus,

$$\sigma_X^2 = E\left[\left(X - \bar{X}\right)^2\right] = \overline{\left(X - \bar{X}\right)^2} = \begin{cases} \sum_i \left(x_i - \bar{X}\right)^2 p_X(x_i) & X \quad \text{discrete} \\ \int_{-\infty}^{\infty} \left(x - \bar{X}\right)^2 f_X(x)\,dx & X \quad \text{continuous} \end{cases}$$

1.3   TRANSFORM METHODS

Different types of transforms are used in science and engineering. In this book we consider two types of transforms: the z-transform of PMFs and the s-transform of PDFs of nonnegative random variables. These transforms are particularly used when random variables take only nonnegative values, which is usually the case in many applications discussed in this book.

1.3.1   The s-Transform

Let $f_X(x)$ be the PDF of the continuous random variable X that takes only nonnegative values; that is, $f_X(x) = 0$ for $x < 0$. The s-transform of $f_X(x)$, denoted by $M_X(s)$, is defined by:

$$M_X(s) = E\left[e^{-sX}\right] = \int_0^{\infty} e^{-sx} f_X(x)\,dx.$$

One important property of an s-transform is that when it is evaluated at the point $s = 0$, its value is equal to 1. That is,

$$M_X(s)\big|_{s=0} = \int_0^{\infty} f_X(x)\,dx = 1.$$

For example, the value of K for which the function $A(s) = K/(s+5)$ is a valid s-transform of a PDF is obtained by setting $A(0) = 1$, which gives:

$$K/5 = 1 \Rightarrow K = 5.$$

1.3.2   Moment-Generating Property of the s-Transform

One of the primary reasons for studying the transform methods is to use them to derive the moments of the different probability distributions. By definition:

$$M_X(s) = \int_0^{\infty} e^{-sx} f_X(x)\,dx.$$

Taking different derivatives of $M_X(s)$ and evaluating them at $s = 0$, we obtain the following results:

$$\frac{d}{ds}M_X(s) = \frac{d}{ds}\int_0^\infty e^{-sx}f_X(x)dx = \int_0^\infty \frac{d}{ds}e^{-sx}f_X(x)dx$$

$$= -\int_0^\infty xe^{-sx}f_X(x)dx,$$

$$\frac{d}{ds}M_X(s)\Big|_{s=0} = -\int_0^\infty xf_X(x)dx$$

$$= -E[X],$$

$$\frac{d^2}{ds^2}M_X(s) = \frac{d}{ds}(-1)\int_{-\infty}^\infty xe^{-sx}f_X(x)dx = \int_0^\infty x^2 e^{-sx}f_X(x)dx,$$

$$\frac{d^2}{ds^2}M_X(s)\Big|_{s=0} = \int_0^\infty x^2 f_X(x)dx$$

$$= E[X^2].$$

In general,

$$\frac{d^n}{ds^n}M_X(s)\Big|_{s=0} = (-1)^n E[X^n].$$

1.3.3   The z-Transform

Let $p_X(x)$ be the PMF of the discrete random variable X. The z-transform of $p_X(x)$, denoted by $G_X(z)$, is defined by:

$$G_X(z) = E[z^X] = \sum_{x=0}^\infty z^x p_X(x).$$

Thus, the PMF $p_X(x)$ is required to take on only nonnegative integers, as we stated earlier. The sum is guaranteed to converge and, therefore, the z-transform exists, when evaluated on or within the unit circle (where $|z| \le 1$). Note that:

$$G_X(1) = \sum_{x=0}^\infty p_X(x) = 1.$$

This means that a valid z-transform of a PMF reduces to unity when evaluated at $z = 1$. However, this is a necessary but not sufficient condition for a function to the z-transform of a PMF. By definition,

$$G_X(z) = \sum_{x=0}^\infty z^x p_X(x)$$

$$= p_X(0) + zp_X(1) + z^2 p_X(2) + z^3 p_X(3) + \dots.$$

This means that $P[X = k] = p_X(k)$ is the coefficient of z^k in the series expansion. Thus, given the z-transform of a PMF, we can uniquely recover the PMF.

The implication of this statement is that not every function of z that has a value of 1 when evaluated at $z = 1$ is a valid z-transform of a PMF. For example, consider the function $A(z) = 2z - 1$. Although $A(1) = 1$, the function contains invalid coefficients in the sense that these coefficients either have negative values or positive values that are greater than one. Thus, for a function of z to be a valid z-transform of a PMF, it must have a value of 1 when evaluated at $z = 1$, and the coefficients of z must be nonnegative numbers that cannot be greater than 1.

The individual terms of the PMF can also be determined as follows:

$$p_X(x) = \frac{1}{x!}\left[\frac{d^x}{dz^x}G_X(z)\right]_{z=0} \qquad x = 0, 1, 2, \ldots.$$

This feature of the z-transform is the reason it is sometimes called the *probability generating function*.

1.3.4 Moment-Generating Property of the z-Transform

As stated earlier, one of the major motivations for studying transform methods is their usefulness in computing the moments of the different random variables. Unfortunately, the moment-generating capability of the z-transform is not as computationally efficient as that of the s-transform.

The moment-generating capability of the z-transform lies in the results obtained from evaluating the derivatives of the transform at $z = 1$. For a discrete random variable X with PMF $p_X(x)$, we have that:

$$G_X(z) = \sum_{x=0}^{\infty} z^x p_X(x),$$

$$\frac{d}{dz}G_X(z) = \frac{d}{dz}\sum_{x=0}^{\infty} z^x p_X(x) = \sum_{x=0}^{\infty} \frac{d}{dz} z^x p_X(x) = \sum_{x=0}^{\infty} x z^{x-1} p_X(x) = \sum_{x=1}^{\infty} x z^{x-1} p_X(x),$$

$$\frac{d}{dz}G_X(z)\Big|_{z=1} = \sum_{x=1}^{\infty} x p_X(x) = \sum_{x=0}^{\infty} x p_X(x) = E[X].$$

Similarly,

$$\frac{d^2}{dz^2}G_X(z) = \frac{d}{dz}\sum_{x=1}^{\infty} x z^{x-1} p_X(x) = \sum_{x=1}^{\infty} x \frac{d}{dz} z^{x-1} p_X(x) = \sum_{x=1}^{\infty} x(x-1)z^{x-2} p_X(x),$$

$$\frac{d^2}{dz^2}G_X(z)\Big|_{z=1} = \sum_{x=1}^{\infty} x(x-1)p_X(x) = \sum_{x=0}^{\infty} x(x-1)p_X(x) = \sum_{x=0}^{\infty} x^2 p_X(x) - \sum_{x=0}^{\infty} x p_X(x)$$

$$= E[X^2] - E[X],$$

$$E[X^2] = \frac{d^2}{dz^2}G_X(z)\Big|_{z=1} + \frac{d}{dz}G_X(z)\Big|_{z=1}.$$

Thus, the variance is obtained as follows:

$$\sigma_X^2 = E[X^2] - (E[X])^2$$

$$= \left[\frac{d^2}{dz^2}G_X(z) + \frac{d}{dz}G_X(z) - \left\{\frac{d}{dz}G_X(z)\right\}^2\right]_{z=1}.$$

1.4   COVARIANCE AND CORRELATION COEFFICIENT

Consider two random variables X and Y with expected values $E[X] = \mu_X$ and $E[Y] = \mu_Y$, respectively, and variances σ_X^2 and σ_Y^2, respectively. The *covariance* of X and Y, which is denoted by $Cov(X, Y)$ or σ_{XY}, is defined by:

$$Cov(X, Y) = \sigma_{XY} = E[(X - \mu_X)(Y - \mu_Y)]$$
$$= E[XY - \mu_Y X - \mu_X Y + \mu_X \mu_Y]$$
$$= E[XY] - \mu_X \mu_Y - \mu_X \mu_Y + \mu_X \mu_Y$$
$$= E[XY] - \mu_X \mu_Y.$$

If X and Y are independent, then $E[XY] = \mu_X \mu_Y$ and $Cov(X, Y) = 0$. However, the converse is not true; that is, if the covariance of X and Y is zero, it does not mean that X and Y are independent random variables. If the covariance of two random variables is zero, we define the two random variables to be *uncorrelated*.

We define the *correlation coefficient* of X and Y, denoted by $\rho(X, Y)$ or ρ_{XY}, as follows:

$$\rho_{XY} = \frac{Cov(X, Y)}{\sqrt{Var(X)Var(Y)}} = \frac{\sigma_{XY}}{\sigma_X \sigma_Y}.$$

The correlation coefficient has the property that:

$$-1 \leq \rho_{XY} \leq 1.$$

1.5   SUMS OF INDEPENDENT RANDOM VARIABLES

Consider two independent continuous random variables X and Y. We are interested in computing the CDF and PDF of their sum $g(X, Y) = U = X + Y$. The random variable S can be used to model the reliability of systems with stand-by connections. In such systems, the component A whose time-to-failure is represented by the random variable X is the primary component, and the component B whose time-to-failure is represented by the random variable Y is the backup component that is brought into operation when the primary component fails. Thus, S represents the time until the system fails, which is the sum of the lifetimes of both components.

Their CDF can be obtained as follows:

$$F_S(s) = P[S \leq s] = P[X + Y \leq s] = \iint_D f_{XY}(x, y)dxdy,$$

where $f_{XY}(x,y)$ is the joint PDF of X and Y and D is the set $D = \{(x,y)|x + y \leq s\}$. Thus,

$$F_S(s) = \int_{-\infty}^{\infty} \int_{-\infty}^{s-y} f_{XY}(x, y)dxdy = \int_{-\infty}^{\infty} \int_{-\infty}^{s-y} f_X(x)f_Y(y)dxdy$$

$$= \int_{-\infty}^{\infty} \left\{ \int_{-\infty}^{s-y} f_X(x)dx \right\} f_Y(y)dy$$

$$= \int_{-\infty}^{\infty} F_X(s-y)f_Y(y)dy.$$

The PDF of S is obtained by differentiating the CDF, as follows:

$$f_S(s) = \frac{d}{ds}F_S(s) = \frac{d}{ds}\int_{-\infty}^{\infty} F_X(s-y)f_Y(y)dy$$

$$= \int_{-\infty}^{\infty} \frac{d}{ds}F_X(s-y)f_Y(y)dy$$

$$= \int_{-\infty}^{\infty} f_X(s-y)f_Y(y)dy,$$

where we have assumed that we can interchange differentiation and integration. The expression on the right-hand side is a well-known result in signal analysis called the *convolution integral*. Thus, we find that the PDF of the sum S of two independent random variables X and Y is the convolution of the PDFs of the two random variables; that is,

$$f_S(s) = f_X(s) * f_Y(s).$$

In general, if S is the sum on n mutually independent random variables X_1, $X_2, \ldots, X_n$ whose PDFs are $f_{X_i}(x), i = 1, 2, \ldots, n$, then we have that:

$$S = X_1 + X_2 + \ldots + X_n,$$
$$f_S(s) = f_{X_1}(s) * f_{X_2}(s) \ldots * f_{X_n}(s).$$

Thus, the s-transform of the PDF of S is given by:

$$M_S(s) = \prod_{i=1}^{n} M_{X_i}(s).$$

1.6 RANDOM SUM OF RANDOM VARIABLES

Let X be a continuous random variable with PDF $f_X(x)$ whose s-transform is $M_X(s)$. We know that if Y is the sum of n independent and identically distributed random variables with the PDF $f_X(x)$, then from the results in the previous section, the s-transform of the PDF of Y is given by:

$$M_Y(s) = [M_X(s)]^n.$$

This result assumes that n is a fixed number. However, there are certain situations when the number of random variables in a sum is itself a random variable. For this case, let N denote a discrete random variable with PMF $p_N(n)$ whose z-transform is $G_N(z)$. Our goal is to find the s-transform of the PDF of Y when the number of random variables is itself a random variable N.

Thus, we consider the sum:

$$Y = X_1 + X_2 + \ldots + X_N,$$

where N has a known PMF, which in turn has a known z-transform. Now, let $N = n$. Then with N fixed at n, we have that:

$$Y|_{N=n} = X_1 + X_2 + \ldots + X_n,$$
$$M_{Y|N}(s|n) = [M_X(s)]^n,$$
$$M_Y(s) = \sum_n p_N(n) M_{Y|N}(s|n) = \sum_n p_N(n)[M_X(s)]^n = G_N(M_X(s)).$$

That is, the s-transform of the PDF of a random sum of independent and identically distributed random variables is the z-transform of the PMF of the number of variables evaluated at the s-transform of the PDF of the constituent random variables. Now, let $u = M_X(s)$. Then,

$$\frac{d}{ds} M_Y(s) = \frac{d}{ds} G_N(M_X(s)) = \left\{ \frac{dG_N(u)}{du} \right\} \left\{ \frac{du}{ds} \right\},$$
$$\frac{d}{ds} M_Y(s)\Big|_{s=0} = \left[\left\{ \frac{dG_N(u)}{du} \right\} \left\{ \frac{du}{ds} \right\} \right]_{s=0}.$$

When $s = 0$, $u|_{s=0} = M_X(0) = 1$. Thus, we obtain:

$$\frac{d}{ds} M_Y(s)\Big|_{s=0} = \left[\left\{ \frac{dG_N(u)}{du} \right\} \left\{ \frac{du}{ds} \right\} \right]_{s=0} = \frac{dG_N(u)}{du}\Big|_{u=1} \frac{dM_X(s)}{ds}\Big|_{s=0},$$
$$-E[Y] = E[N](-E[X]) = -E[N]E[X],$$
$$E[Y] = E[N]E[X].$$

Also,

$$\frac{d^2}{ds^2} M_Y(s) = \frac{d}{ds} \left[\left\{ \frac{dG_N(u)}{du} \right\} \left\{ \frac{du}{ds} \right\} \right] = \left\{ \frac{du}{ds} \right\} \frac{d}{ds} \left\{ \frac{dG_N(u)}{du} \right\} + \left\{ \frac{dG_N(u)}{du} \right\} \left\{ \frac{d^2u}{ds^2} \right\}$$
$$= \left\{ \frac{du}{ds} \right\}^2 \left\{ \frac{d^2G_N(u)}{du^2} \right\} + \left\{ \frac{dG_N(u)}{du} \right\} \left\{ \frac{d^2u}{ds^2} \right\},$$

$$\frac{d^2}{ds^2}M_Y(s)\Big|_{s=0} = E[Y^2] = \left[\left\{\frac{du}{ds}\right\}^2\left\{\frac{d^2G_N(u)}{du^2}\right\} + \left\{\frac{dG_N(u)}{du}\right\}\left\{\frac{d^2u}{ds^2}\right\}\right]_{s=0;\,u=1}$$
$$= \{-E[X]\}^2\{E[N^2]-E[N]\} + E[N]E[X^2]$$
$$= E[N^2]\{E[X]\}^2 + E[N]E[X^2] - E[N]\{E[X]\}^2.$$

The variance of Y is given by:

$$\sigma_Y^2 = E[Y^2] - (E[Y])^2$$
$$= E[N^2]\{E[X]\}^2 + E[N]E[X^2] - E[N]\{E[X]\}^2$$
$$\quad - (E[N]E[X])^2$$
$$= E[N]\{E[X^2] - \{E[X]\}^2\} + (E[X])^2\{E[N^2] - (E[N])^2\}$$
$$= E[N]\sigma_X^2 + (E[X])^2\sigma_N^2.$$

If X is also a discrete random variable, then we obtain:

$$G_Y(z) = G_N(G_X(z)),$$

and the results for $E[Y]$ and σ_Y^2 still hold.

1.7 SOME PROBABILITY DISTRIBUTIONS

Random variables with special probability distributions are encountered in different fields of science and engineering. In this section we describe some of these distributions, including their expected values, variances, and s-transforms (or z-transforms, as the case may be).

1.7.1 The Bernoulli Distribution

A Bernoulli trial is an experiment that results in two outcomes: *success* and *failure*. One example of a Bernoulli trial is the coin-tossing experiment, which results in heads or tails. In a Bernoulli trial we define the probability of success and probability of failure as follows:

$$P[\text{success}] = p \qquad 0 \le p \le 1$$
$$P[\text{failure}] = 1 - p$$

Let us associate the events of the Bernoulli trial with a random variable X such that when the outcome of the trial is a success, we define $X = 1$, and when the outcome is a failure, we define $X = 0$. The random variable X is called a Bernoulli random variable, and its PMF is given by:

$$P_X(x) = \begin{cases} 1-p & x=0 \\ p & x=1 \end{cases}$$

An alternative way to define the PMF of X is as follows:

$$p_X(x) = p^x(1-p)^{1-x} \quad x = 0, 1.$$

The CDF is given by:

$$F_X(x) = \begin{cases} 0 & x < 0 \\ 1-p & 0 \le x < 1 \\ 1 & x \ge 1 \end{cases}$$

The expected value of X is given by:

$$E[X] = 0(1-p) + 1(p) = p.$$

Similarly, the second moment of X is given by:

$$E[X^2] = 0^2(1-p) + 1^2(p) = p.$$

Thus, the variance of X is given by:

$$\sigma_X^2 = E[X^2] - \{E[X]\}^2 = p - p^2 = p(1-p).$$

The z-transform of the PMF is given by:

$$G_X(z) = \sum_{x=0}^{\infty} z^x p_X(x) = \sum_{x=0}^{1} z^x p_X(x) = z^0(1-p) + z^1 p = 1 - p + zp.$$

1.7.2   The Binomial Distribution

Suppose we conduct n independent Bernoulli trials and we represent the number of successes in those n trials by the random variable $X(n)$. Then, $X(n)$ is defined as a binomial random variable with parameters (n, p). The PMF of a random variable, $X(n)$, with parameters (n, p) is given by:

$$p_{X(n)}(x) = \binom{n}{x} p^x(1-p)^{n-x} \quad x = 0, 1, 2, \ldots, n.$$

The binomial coefficient, $\binom{n}{x}$, represents the number of ways of arranging x successes and $n - x$ failures.

The CDF, mean and variance of $X(n)$, and the z-transform of its PMF are given by:

$$F_{X(n)}(x) = P[X(n) \leq x] = \sum_{k=0}^{x} \binom{n}{k} p^k (1-p)^{n-k},$$

$$E[X(n)] = np,$$

$$E[X^2(n)] = n(n-1)p^2 + np,$$

$$\sigma_{X(n)}^2 = E[X^2(n)] - \{E[X(n)]\}^2 = np(1-p),$$

$$G_{X(n)}(z) = (zp + 1 - p)^n.$$

1.7.3 The Geometric Distribution

The geometric random variable is used to describe the number of independent Bernoulli trials until the first success occurs. Let X be a random variable that denotes the number of Bernoulli trials until the first success. If the first success occurs on the xth trial, then we know that the first $x - 1$ trials resulted in failures. Thus, the PMF of a geometric random variable, X, is given by:

$$p_X(x) = p(1-p)^{x-1} \quad x = 1, 2, 3, \dots.$$

The CDF, mean, and variance of X and the z-transform of its PMF are given by:

$$F_X(x) = P[X \leq x] = 1 - (1-p)^x,$$

$$E[X] = 1/p,$$

$$E[X^2] = \frac{2-p}{p^2},$$

$$\sigma_X^2 = E[X^2] - \{E[X]\}^2 = \frac{1-p}{p^2},$$

$$G_X(z) = \frac{zp}{1 - z(1-p)}.$$

1.7.4 The Pascal Distribution

The Pascal random variable is an extension of the geometric random variable. A Pascal random variable of order k describes the number of trials until the kth success, which is why it is sometimes called the "kth-order interarrival time for a Bernoulli process." The Pascal distribution is also called the *negative binomial distribution*.

Let X_k be a kth-order Pascal random variable. Then its PMF is given by:

$$p_{X_k}(n) = \binom{n-1}{k-1} p^k (1-p)^{n-k} \quad k = 1, 2, \dots; n = k, k+1, \dots.$$

The CDF, mean, and variance of X_k and the z-transform of its PMF are given by:

$$F_{x_k}(x) = P[X_k \le x] = \sum_{n=k}^{x} \binom{n-1}{k-1} p^k (1-p)^{n-k},$$

$$E(X_k) = k/p,$$

$$E[X_k^2] = \frac{k^2 + k(1-p)}{p^2},$$

$$\sigma_{X_k}^2 = E[X_k^2] - \{E[X_k]\}^2 = \frac{k(1-p)}{p^2},$$

$$G_{X_k}(z) = \left[\frac{zp}{1-z(1-p)} \right]^k.$$

1.7.5 The Poisson Distribution

A discrete random variable K is called a Poisson random variable with parameter λ, where $\lambda > 0$, if its PMF is given by:

$$p_K(k) = \frac{\lambda^k}{k!} e^{-\lambda} \quad k = 0, 1, 2, \ldots.$$

The CDF, mean, and variance of K and the z-transform of its PMF are given by:

$$F_K(k) = P[K \le k] = \sum_{r=0}^{k} \frac{\lambda^r}{r!} e^{-\lambda},$$

$$E[K] = \lambda,$$

$$E[K^2] = \lambda^2 + \lambda,$$

$$\sigma_K^2 = E[K^2] - \{E[K]\}^2 = \lambda,$$

$$G_K(z) = e^{\lambda(z-1)}.$$

1.7.6 The Exponential Distribution

A continuous random variable X is defined to be an exponential random variable (or X has an exponential distribution) if for some parameter $\lambda > 0$ its PDF is given by:

$$f_X(x) = \begin{cases} \lambda e^{-\lambda x} & x \ge 0 \\ 0 & x < 0 \end{cases}$$

The CDF, mean, and variance of X and the s-transform of its PDF are given by:

$$F_X(x) = P[X \le x] = 1 - e^{-\lambda x},$$
$$E[X] = 1/\lambda,$$
$$E[X^2] = 2/\lambda^2,$$
$$\sigma_X^2 = E[X^2] - \{E[X]\}^2 = 1/\lambda^2,$$
$$M_X(s) = \frac{\lambda}{s+\lambda}.$$

1.7.7 The Erlang Distribution

The Erlang distribution is a generalization of the exponential distribution. While the exponential random variable describes the time between adjacent events, the Erlang random variable describes the time interval between any event and the kth following event. A random variable is referred to as a kth-order Erlang (or Erlang-k) random variable with parameter λ if its PDF is given by:

$$f_{X_k}(x) = \begin{cases} \dfrac{\lambda^k x^{k-1} e^{-\lambda x}}{(k-1)!} & k = 1, 2, 3, \ldots; x \ge 0 \\ 0 & x < 0 \end{cases}$$

The CDF, mean, and variance of X_k and the s-transform of its PDF are given by

$$F_{X_k}(x) = P[X_k \le x] = 1 - \sum_{j=0}^{k-1} \frac{(\lambda x)^j e^{-\lambda x}}{j!},$$
$$E[X_k] = k/\lambda,$$
$$E[X_k^2] = \frac{k(k+1)}{\lambda^2},$$
$$\sigma_{X_k}^2 = E[X_k^2] - \{E[X_k]\}^2 = \frac{k}{\lambda^2},$$
$$M_{X_k}(s) = \left[\frac{\lambda}{s+\lambda}\right]^k.$$

1.7.8 The Uniform Distribution

A continuous random variable X is said to have a uniform distribution over the interval $[a, b]$ if its PDF is given by:

$$f_X(x) = \begin{cases} \dfrac{1}{b-a} & a \le x \le b \\ 0 & \text{otherwise} \end{cases}$$

The CDF, mean, and variance of X and the s-transform of its PDF are given by:

$$F_X(x) = P[X \leq x] = \begin{cases} 0 & x < a \\ \dfrac{x-a}{b-a} & a \leq x < b, \\ 1 & x \geq b \end{cases}$$

$$E[X] = \frac{b+a}{2},$$

$$E[X^2] = \frac{b^2 + ab + a^2}{3},$$

$$\sigma_X^2 = E[X^2] - \{E[X]\}^2 = \frac{(b-a)^2}{12},$$

$$M_X(s) = \frac{e^{-as} - e^{-bs}}{s(b-a)}$$

1.7.9 The Hyperexponential Distribution

The Erlang distribution belongs to a class of distributions that are said to have a *phase-type* distribution. This arises from the fact that the Erlang distribution is the sum of independent exponential distributions. Thus, an Erlang random variable can be thought of as the time to go through a sequence of phases or stages, each of which requires an exponentially distributed length of time. For example, since an Erlang-k random variable X_k is the sum of k exponentially distributed random variables X with mean $1/\mu$, and we can visualize X_k as the time it takes to complete a task that must go through k stages, where the time the task spends at each stage is X. Thus, we can represent the time to complete that task by the series of stages shown in Figure 1.1.

The hyperexponential distribution is another type of the phase-type distribution. The random variable H_k is used to model a process where an item can choose one of k branches. The probability that it chooses branch i is α_i, $i = 1$, $2, \ldots, k$. The time it takes the item to traverse branch i is exponentially distributed with a mean of $1/\mu_i$. Thus, the PDF of H_k is given by:

$$f_{H_k}(x) = \sum_{i=1}^{k} \alpha_i \mu_i e^{-\mu_i x}, \quad x \geq 0$$

$$\sum_{i=1}^{k} \alpha_i = 1.$$

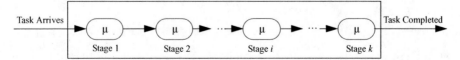

Figure 1.1   Graphical representation of the Erlang-k random variable.

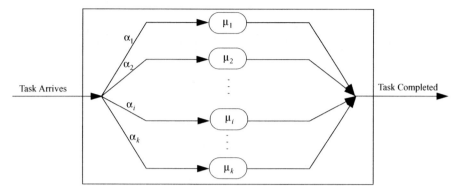

Figure 1.2   Graphical representation of H_k.

The random variable can be visualized as in Figure 1.2.

The mean, second moment, and s-transform of H_k are given by:

$$E[H_k] = \sum_{i=1}^{k} \frac{\alpha_i}{\mu_i},$$

$$E[H_k^2] = 2\sum_{i=1}^{k} \frac{\alpha_i}{\mu_i^2},$$

$$M_{H_k}(s) = \sum_{i=1}^{k} \frac{\alpha_i \mu_i}{s + \mu_i}.$$

1.7.10   The Coxian Distribution

The Coxian distribution is the third member of the phase-type distribution. A random variable C_k has a Coxian distribution of order k if it has to go through up to at most k stages, each of which has an exponential distribution. The random variable is popularly used to approximate general nonnegative distributions with exponential phases. The mean time spent at stage i is $1/\mu_i$, $i = 1$, $2, \ldots, k$. A task arrives at stage 1; it may choose to receive some service at stage 1 with probability β_1 or leave the system with probability $\alpha_1 = 1 - \beta_1$. Given that it receives service at stage 1, the task may leave the system with probability α_2 or proceed to receive further service at stage 2 with probability $\beta_2 = 1 - \alpha_2$. This process continues until the task reaches stage k, where it finally leaves the system after service. The graphical representation of the process is shown in Figure 1.3.

The probability B_i of advancing to the ith stage to receive service is given by:

$$B_i = \prod_{j=1}^{i} \beta_j \quad i = 1, 2, \ldots, k.$$

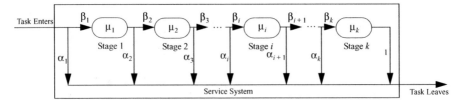

Figure 1.3   Graphical representation of C_k.

Thus, the probability L_i of leaving the system after the ith stage is given by:

$$L_i = \begin{cases} \alpha_{i+1} B_i & i = 1, 2, \dots, k-1 \\ B_k & i = k \end{cases}$$

If we define the PDF of the service time X_i at stage i by $f_{X_i}(x)$, $x \geq 0$, then we note that C_k takes on the following values with the associated probabilities:

$$L_1 = P[C_k = X_1],$$
$$L_2 = P[C_k = X_1 + X_2],$$
$$L_3 = P[C_k = X_1 + X_2 + X_3] = P\left[C_k = \sum_{i=1}^{3} X_i\right],$$
$$\vdots$$
$$L_k = P\left[C_k = \sum_{i=1}^{k} X_i\right].$$

Also, let $g_i(x)$ denote the PDF of the sum of random variables $X_1 + X_2 + \dots + X_i$. Then, we know that $g_i(x)$ is the convolution of the PDFs of the X_i, that is,

$$g_i(x) = f_{X_1}(x) * f_{X_2}(x) * \dots * f_{X_i}(x).$$

Therefore, the s-transform of $g_i(x)$ is:

$$M_{G_i}(s) = M_{X_1}(s) M_{X_2}(s) \dots M_{X_i}(s) = \prod_{j=1}^{i} \frac{\mu_j}{s + \mu_j}.$$

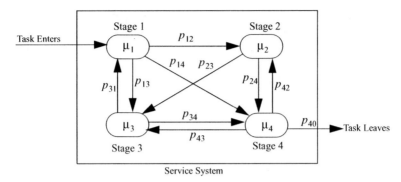

Figure 1.4 Graphical representation of phase-type distribution.

This means that the s-transform of C_k is given by:

$$M_{C_k}(s) = L_1 M_{G_1}(s) + L_2 M_{G_2}(s) + \ldots + L_k M_{G_k}(s) = \sum_{i=1}^{k} L_i M_{G_i}(s)$$

$$= \sum_{i=1}^{k} L_i \left\{ \prod_{j=1}^{i} \frac{\mu_j}{s + \mu_j} \right\},$$

The mean and second moment of C_k are given by:

$$E[C_k] = \sum_{i=1}^{k} L_i \left\{ \prod_{j=1}^{i} \frac{1}{\mu_j} \right\},$$

$$E[C_k^2] = 2 \sum_{i=1}^{k} L_i \left\{ \sum_{j=1}^{i} \left[\frac{1}{\mu_j^2} + \sum_{\substack{l=1 \\ l \neq j}}^{i} \frac{1}{\mu_l \mu_j} \right] \right\}.$$

1.7.11 The General Phase-Type Distribution

The three types of phase-type distributions (Erlang, hyperexponential, and Coxian) are represented by feedforward networks of stages. A more general type of the phase-type distribution allows both feedforward and feedback relationships among the stages. This type is simply called the phase-type distribution. An example is illustrated in Figure 1.4.

This distribution is characterized by both the mean service time $1/\mu_i$ at stage i and a transition probability matrix that defines the probability p_{ij} that a task that has completed service at stage i goes next to stage j. The details of this particular type of distribution are very involved and will not be discussed here.

1.7.12 Normal Distribution

A continuous random variable X is defined to be a normal random variable with parameters μ_X and σ_X^2 if its PDF is given by:

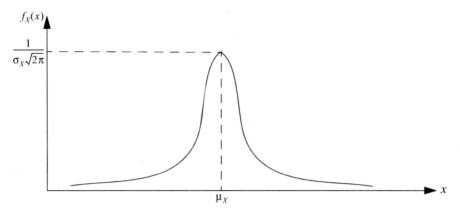

Figure 1.5   PDF of the normal random variable.

$$f_X(x) = \frac{1}{\sqrt{2\pi\sigma_X^2}} e^{-(x-\mu_X)^2/2\sigma_X^2} = \frac{1}{\sqrt{2\pi\sigma_X^2}} e^{-\frac{1}{2}\left(\frac{x-\mu_X}{\sigma_X}\right)^2} \qquad -\infty < x < \infty.$$

The PDF is a bell-shaped curve that is symmetric about μ_X, which is the mean of X. The parameter σ_X^2 is the variance. Figure 1.5 illustrates the shape of the PDF.

The CDF of X is given by:

$$F_X(x) = P[X \le x] = \frac{1}{\sigma_X\sqrt{2\pi}} \int_{-\infty}^{x} e^{-(u-\mu_X)^2/2\sigma_X^2} \, du.$$

The normal random variable X with parameters μ_X and σ_X^2 is usually designated $X = N(\mu_X, \sigma_X^2)$. The special case of zero mean and unit variance (i.e., $\mu_X = 0$ and $\sigma_X^2 = 1$) is designated $X = N(0, 1)$ and is called the *standard normal random variable*. Let $y = (u - \mu_X)/\sigma_X$. Then, $du = \sigma_X dy$ and the CDF of X becomes:

$$F_X(x) = \frac{1}{\sqrt{2\pi}} \int_{-\infty}^{(x-\mu_X)/\sigma_X} e^{-y^2/2} dy.$$

Thus, with the above transformation, X becomes a standard normal random variable. The above integral cannot be evaluated in closed form. It is usually evaluated numerically through the function $\Phi(x)$, which is defined as follows:

$$\Phi(x) = \frac{1}{\sqrt{2\pi}} \int_{-\infty}^{x} e^{-y^2/2} dy.$$

Thus, the CDF of X is given by

$$F_X(x) = \frac{1}{\sqrt{2\pi}} \int_{-\infty}^{(x-\mu_X)/\sigma_X} e^{-y^2/2} dy = \Phi\left(\frac{x-\mu_X}{\sigma_X}\right).$$

The values of $\Phi(x)$ are sometimes given for nonnegative values of x. For negative values of x, $\Phi(x)$ can be obtained from the following relationship:

$$\Phi(-x) = 1 - \Phi(x).$$

Values of $\Phi(x)$ are given in standard books on probability, such as Ibe (2005).

1.8 LIMIT THEOREMS

In this section we discuss two fundamental theorems in probability. These are the law of large numbers, which is regarded as the first fundamental theorem, and the central limit theorem, which is regarded as the second fundamental theorem. We begin the discussion with the Markov and Chebyshev inequalities that enable us to prove these theorems.

1.8.1 Markov Inequality

The Markov inequality applies to random variables that take only nonnegative values. It can be stated as follows:

Proposition 1.1: If X is a random variable that takes only nonnegative values, then for any $a > 0$,

$$P[X \ge a] \le \frac{E[X]}{a}.$$

Proof: We consider only the case when X is a continuous random variable. Thus,

$$
\begin{aligned}
E[X] &= \int_0^\infty x f_X(x) dx = \int_0^a x f_X(x) dx + \int_a^\infty x f_X(x) dx \\
&\ge \int_a^\infty x f_X(x) dx \\
&\ge \int_a^\infty a f_X(x) dx \\
&= a \int_a^\infty f_X(x) dx \\
&= a P[X \ge a],
\end{aligned}
$$

and the result follows.

1.8.2 Chebyshev Inequality

The Chebyshev inequality enables us to obtain bounds on probability when both the mean and variance of a random variable are known. The inequality can be stated as follows:

Proposition 1.2: Let X be a random variable with mean μ and variance σ^2. Then, for any $b > 0$,

$$P[|X - \mu| \geq b] \leq \frac{\sigma^2}{b^2}.$$

Proof: Since $(X - \mu)^2$ is a nonnegative random variable, we can invoke the Markov inequality, with $a = b^2$, to obtain:

$$P\left[(X - \mu)^2 \geq b^2\right] \leq \frac{E\left[(X - \mu)^2\right]}{b^2}.$$

Since $(X - \mu)^2 \geq b^2$ if and only if $|X - \mu| \geq b$, the preceding inequality is equivalent to:

$$P[|X - \mu| \geq b] \leq \frac{E\left[(X - \mu)^2\right]}{b^2} = \frac{\sigma^2}{b^2},$$

which completes the proof.

1.8.3 Law of Large Numbers

There are two laws of large numbers that deal with the limiting behavior of random sequences. One is called the "weak" law of large numbers and the other is called the "strong" law of large numbers. We will discuss only the weak law of large numbers.

Proposition 1.3: Let $X_1, X_2, \ldots, X_n$ be a sequence of mutually independent and identically distributed random variables, and let their mean be $E[X_k] = \mu < \infty$. Similarly, let their variance be $\sigma_{X_k}^2 = \sigma^2 < \infty$. Let S_n denote the sum of the n random variables, that is,

$$S_n = X_1 + X_2 + \ldots + X_n.$$

Then the weak law of large numbers states that for any $\varepsilon > 0$,

$$\lim_{n \to \infty} P\left[\left|\frac{S_n}{n} - \mu\right| \geq \varepsilon\right] \to 0.$$

Equivalently,

$$\lim_{n \to \infty} P\left[\left|\frac{S_n}{n} - \mu\right| < \varepsilon\right] \to 1.$$

Proof: Since $X_1, X_2, \ldots, X_n$ are independent and have the same distribution, we have that:

$$\text{Var}(S_n) = n\sigma^2,$$

$$\text{Var}\left(\frac{S_n}{n}\right) = \frac{n\sigma^2}{n^2} = \frac{\sigma^2}{n},$$

$$E\left[\frac{S_n}{n}\right] = \frac{n\mu}{n} = \mu.$$

From Chebyshev inequality, for $\varepsilon > 0$, we have that:

$$P\left[\left|\frac{S_n}{n} - \mu\right| \geq \varepsilon\right] \leq \frac{\sigma^2}{n\varepsilon^2}.$$

Thus, for a fixed ε,

$$P\left[\left|\frac{S_n}{n} - \mu\right| \geq \varepsilon\right] \to 0$$

as $n \to \infty$, which completes the proof.

1.8.4 The Central Limit Theorem

The central limit theorem provides an approximation to the behavior of sums of random variables. The theorem states that as the number of independent and identically distributed random variables with finite mean and finite variance increases, the distribution of their sum becomes increasingly normal regardless of the form of the distribution of the random variables. More formally, let $X_1, X_2, \ldots, X_n$ be a sequence of mutually independent and identically distributed random variables, each of which has a finite mean μ_X and a finite variance σ_X^2. Let S_n be defined as follows:

$$S_n = X_1 + X_2 + \ldots + X_n.$$

Now,

$$E[S_n] = n\mu_X,$$

$$\sigma_{S_n}^2 = n\sigma_X^2.$$

Converting S_n to standard normal random variable (i.e., zero mean and variance = 1) we obtain:

$$Y_n = \frac{S_n - \overline{S}_n}{\sigma_{S_n}} = \frac{S_n - n\mu_X}{\sqrt{n\sigma_X^2}} = \frac{S_n - n\mu_X}{\sigma_X \sqrt{n}}.$$

The central limit theorem states that if $F_{Y_n}(y)$ is the CDF of Y_n, then:

$$\lim_{n \to \infty} F_{Y_n}(y) = \lim_{n \to \infty} P[Y_n \le y] = \frac{1}{\sqrt{2\pi}} \int_{-\infty}^{y} e^{-u^2/2} du = \Phi(y).$$

This means that $\lim_{n \to \infty} Y_n \sim N(0, 1)$. Thus, one of the important roles that the normal distribution plays in statistics is its usefulness as an approximation of other probability distribution functions.

An alternate statement of the theorem is that in the limit as n becomes very large,

$$\tilde{S}_n = \frac{S_n}{\sigma_{S_n}} = \frac{S_n}{\sigma_X \sqrt{n}} = \frac{X_1 + X_2 + \ldots + X_n}{\sigma_X \sqrt{n}}$$

is a normal random variable with unit variance.

1.9 PROBLEMS

1.1 A sequence of Bernoulli trials consists of choosing seven components at random from a batch of components. A selected component is classified as either defective or nondefective. A nondefective component is considered to be a success, while a defective component is considered to be a failure. If the probability that a selected component is nondefective is 0.8, what is the probability of exactly three successes?

1.2 The probability that a patient recovers from a rare blood disease is 0.3. If 15 people are known to have contracted this disease, find the following probabilities:

a. At least 10 survive.

b. From three to eight survive.

c. Exactly six survive.

1.3 A sequence of Bernoulli trials consists of choosing components at random from a batch of components. A selected component is classified as either defective or nondefective. A nondefective component is considered to be a success, while a defective component is considered to be a failure. If the probability that a selected component is nondefective is 0.8, determine the probabilities of the following events:

a. The first success occurs on the fifth trial.

b. The third success occurs on the eighth trial.

c. There are two successes by the fourth trial, there are four successes by the 10th trial, and there are 10 successes by the 18th trial.

1.4 A lady invites 12 people for dinner at her house. Unfortunately the dining table can only seat six people. Her plan is that if six or fewer guests come, then they will be seated at the table (i.e., they will have a sit-down dinner); otherwise, she will set up a buffet-style meal. The probability that each invited guest will come to dinner is 0.4, and each guest's decision is independent of other guests' decisions. Determine the following:

a. The probability that she has a sit-down dinner

b. The probability that she has a buffet-style dinner

c. The probability that there are at most three guests

1.5 A Girl Scout troop sells cookies from house to house. One of the parents of the girls figured out that the probability that they sell a set of packs of cookies at any house they visit is 0.4, where it is assumed that they sell exactly one set to each house that buys their cookies.

a. What is the probability that the first house where they make their first sale is the fifth house they visit?

b. Given that they visited 10 houses on a particular day, what is the probability that they sold exactly six sets of cookie packs?

c. What is the probability that on a particular day the third set of cookie packs is sold at the seventh house that the girls visit?

1.6 Students arrive for a lab experiment according to a Poisson process with a rate of 12 students per hour. However, the lab attendant opens the door to the lab when at least four students are waiting at the door. What is the probability that the waiting time of the first student to arrive exceeds 20 min? (By waiting time we mean the time that elapses from when a student arrives until the door is opened by the lab attendant.)

1.7 Cars arrive at a gas station according to a Poisson process at an average rate of 12 cars per hour. The station has only one attendant. If the attendant decides to take a 2-min coffee break when there were no cars at the station, what is the probability that one or more cars will be waiting when he comes back from the break, given that any car that arrives when he is on coffee break waits for him to get back?

1.8 An insurance company pays out claims on its life insurance policies in accordance with a Poisson process with an average rate of five claims per week. If the amount of money paid on each policy is uniformly distributed between $2000 and $10,000, what is the mean of the total amount of money that the company pays out in a 4-week period?

1.9 Three customers A, B, and C simultaneously arrive at a bank with two tellers on duty. The two tellers were idle when the three customers arrived, and A goes directly to one teller, B goes to the other teller, and C waits until either A or B leaves before she can begin receiving service. If the service times provided by the tellers are exponentially distributed with a mean of 4 min, what is the probability that customer A is still in the bank after the other two customers leave?

1.10 A five-motor machine can operate properly if at least three of the five motors are functioning. If the lifetime X of each motor has the PDF $f_X(x) = \lambda e^{-\lambda x}, x \geq 0, \lambda > 0$, and if the lifetimes of the motors are independent, what is the mean of the random variable Y, the time until the machine fails?

2

OVERVIEW OF STOCHASTIC PROCESSES

2.1 INTRODUCTION

Stochastic processes deal with the dynamics of probability theory. The concept of stochastic processes enlarges the random variable concept to include time. Thus, instead of thinking of a random variable X that maps an event $w \in \Omega$, where Ω is the sample space, to some number $X(w)$, we think of how the random variable maps the event to different numbers at different times. This implies that instead of the number $X(w)$ we deal with $X(t, w)$, where $t \in T$ and T is called the *parameter set* of the process and is usually a set of times.

Stochastic processes are widely encountered in such fields as communications, control, management science, and time series analysis. Examples of stochastic processes include the population growth, the failure of equipment, the price of a given stock over time, and the number of calls that arrive at a switchboard.

If we fix the sample point w, we obtain $X(t)$, which is some real function of time; and for each w, we have a different function $X(t)$. Thus, $X(t, w)$ can be viewed as a collection of time functions, one for each sample point w. On the other hand, if we fix t, we have a function $X(w)$ that depends only on w and thus is a random variable. Thus, a stochastic process becomes a random variable when time is fixed at some particular value. With many values of t we obtain a collection of random variables. Thus, we can define a stochastic process as a family of random variables $\{X(t, w) | t \in T, w \in \Omega\}$ defined over a

Fundamentals of Stochastic Networks, First Edition. Oliver C. Ibe.
© 2011 John Wiley & Sons, Inc. Published 2011 by John Wiley & Sons, Inc.

given probability space and indexed by the time parameter t. A stochastic process is also called a *random process*. Thus, we will use the terms "stochastic process" and "random process" interchangeably.

2.2 CLASSIFICATION OF STOCHASTIC PROCESSES

A stochastic process can be classified according to the nature of the time parameter and the values that $X(t, w)$ can take on. As discussed earlier, T is called the parameter set of the random process. If T is an interval of real numbers and hence is continuous, the process is called a *continuous-time* stochastic process. Similarly, if T is a countable set and hence is discrete, the process is called a *discrete-time* random process. A discrete-time stochastic process is also called a *random sequence*, which is denoted by $\{X[n]|n = 1, 2, \ldots\}$.

The values that $X(t, w)$ assumes are called the *states* of the stochastic process. The set of all possible values of $X(t, w)$ forms the *state space*, S, of the stochastic process. If S is continuous, the process is called a *continuous-state* stochastic process. Similarly, if S is discrete, the process is called a *discrete-state* stochastic process. In the remainder of this book we denote a stochastic process by $X(t)$, suppressing the parameter w.

2.3 STATIONARY RANDOM PROCESSES

There are several ways to define a stationary random process. At a high level, it is a process whose statistical properties do not vary with time. In this book we consider only two types of stationary processes. These are the *strict-sense stationary* processes and the *wide-sense stationary* (WSS) processes.

2.3.1 Strict-Sense Stationary Processes

A random process is defined to be a strict-sense stationary process if its cumulative distribution function (CDF) is invariant to a shift in the time origin. This means that the process $X(t)$ with the CDF $F_X(x_1, x_2, \ldots, x_n; t_1, t_2, \ldots, t_n)$ is a strict-sense stationary process if its CDF is identical to that of $X(t + \varepsilon)$ for any arbitrary ε. Thus, we have that being a strict-sense stationary process implies that for any arbitrary ε,

$$F_X(x_1, x_2, \ldots, x_n; t_1, t_2, \ldots, t_n) = F_X(x_1, x_2, \ldots, x_n; t_1 + \varepsilon, t_2 + \varepsilon, \ldots, t_n + \varepsilon) \quad \text{for all } n.$$

When the CDF is differentiable, the equivalent condition for strict-sense stationarity is that the probability density function (PDF) is invariant to a shift in the time origin; that is,

$$f_X(x_1, x_2, \ldots, x_n; t_1, t_2, \ldots, t_n) = f_X(x_1, x_2, \ldots, x_n; t_1 + \varepsilon, t_2 + \varepsilon, \ldots, t_n + \varepsilon) \quad \text{for all } n.$$

If $X(t)$ is a strict-sense stationary process, then the CDF $F_{X_1 X_2}(x_1, x_2; t_1, t_1 + \tau)$ does not depend on t but it may depend on τ. Thus, if $t_2 = t_1 + \tau$, then $F_{X_1 X_2}(x_1, x_2; t_1, t_2)$ may depend on $t_2 - t_1$, but not on t_1 and t_2 individually. This means that if $X(t)$ is a strict-sense stationary process, then the autocorrelation and autocovariance functions do not depend on t. Thus, we have that for all $\tau \in T$:

$$\mu_X(t) = \mu_X(0),$$
$$R_{XX}(t, t + \tau) = R_{XX}(0, \tau),$$
$$C_{XX}(t, t + \tau) = C_{XX}(0, \tau).$$

If the condition $\mu_X(t) = \mu_X(0)$ holds for all t, the mean is constant and denoted by μ_X. Similarly, if the equation $R_{XX}(t, t + \tau)$ does not depend on t but is a function of τ, we write $R_{XX}(0, \tau) = R_{XX}(\tau)$. Finally, whenever the condition $C_{XX}(t, t + \tau) = C_{XX}(0, \tau)$ holds for all t, we write $C_{XX}(0, \tau) = C_{XX}(\tau)$.

2.3.2   WSS Processes

Many practical problems that we encounter require that we deal with only the mean and autocorrelation function of a random process. Solutions to these problems are simplified if these quantities do not depend on absolute time. Random processes in which the mean and autocorrelation function do not depend on absolute time are called WSS processes. Thus, for a WSS process $X(t)$,

$$E[X(t)] = \mu_X \quad \text{(constant)}$$
$$R_{XX}(t, t + \tau) = R_{XX}(\tau)$$

Note that a strict-sense stationary process is also a WSS process. However, in general the converse is not true; that is, a WSS process is not necessarily stationary in the strict sense.

2.4   COUNTING PROCESSES

A random process $\{X(t) | t \geq 0\}$ is called a counting process if $X(t)$ represents the total number of "events" that have occurred in the interval $[0, t)$. An example of a counting process is the number of customers that arrive at a bank from the time the bank opens its doors for business until some time t. A counting process satisfies the following conditions:

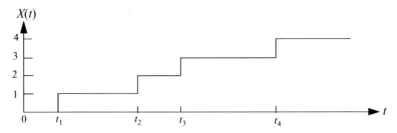

Figure 2.1 Sample function of a counting process.

1. $X(t) \geq 0$, which means that it has nonnegative values.
2. $X(0) = 0$, which means that the counting of events begins at time 0.
3. $X(t)$ is integer valued.
4. If $s < t$, then $X(s) \leq X(t)$, which means that it is a nondecreasing function of time.
5. $X(t) - X(s)$ represents the number of events that have occurred in the interval $[s, t]$.

Figure 2.1 represents a sample path of a counting process. The first event occurs at time t_1, and subsequent events occur at times t_2, t_3, and t_4. Thus, the number of events that occur in the interval $[0, t_4]$ is 4.

2.5   INDEPENDENT INCREMENT PROCESSES

A counting process is defined to be an independent increment process if the number of events that occur in disjoint time intervals is an independent random variable. For example, in Figure 2.1, consider the two nonoverlapping (i.e., disjoint) time intervals $[0, t_1]$ and $[t_2, t_4]$. If the number of events occurring in one interval is independent of the number of events that occur in the other, then the process is an independent increment process. Thus, $X(t)$ is an independent increment process if for every set of time instants $t_0 = 0 < t_1 < t_2 < \ldots < t_n$ the increments $X(t_1) - X(t_0)$, $X(t_2) - X(t_1)$, $\ldots$, $X(t_n) - X(t_{n-1})$ are mutually independent random variables.

2.6   STATIONARY INCREMENT PROCESS

A counting process $X(t)$ is defined to possess stationary increments if for every set of time instants $t_0 = 0 < t_1 < t_2 < \ldots < t_n$ the increments $X(t_1) - X(t_0)$, $X(t_2) - X(t_1)$, $\ldots$, $X(t_n) - X(t_{n-1})$ are identically distributed. In general, the mean of an independent increment process $X(t)$ with stationary increments has the form:

$$E[X(t)] = mt,$$

where the constant m is the value of the mean at time $t = 1$. That is, $m = E[X(1)]$. Similarly, the variance of an independent increment process $X(t)$ with stationary increments has the form:

$$\mathrm{Var}[X(t)] = \sigma^2 t,$$

where the constant σ^2 is the value of the variance at time $t = 1$; that is, $\sigma^2 = \mathrm{Var}[X(1)]$.

2.7 POISSON PROCESSES

Poisson processes are widely used to model arrivals (or occurrence of events) in a system. For example, they are used to model the arrival of telephone calls at a switchboard, the arrival of customers' orders at a service facility, and the random failures of equipment. There are two ways to define a Poisson process. The first definition of the process is that it is a counting process $X(t)$ in which the number of events in any interval of length t has a Poisson distribution with mean λt. Thus, for all $s, t > 0$,

$$P[X(s+t) - X(s) = n] = \frac{(\lambda t)^n}{n!} e^{-\lambda t} \quad n = 0, 1, 2, \dots .$$

The second way to define the Poisson process $X(t)$ is that it is a counting process with stationary and independent increments such that for a rate $\lambda > 0$, the following conditions hold:

1. $P[X(t + \Delta t) - X(t) = 1] = \lambda \Delta t + o(\Delta t)$, which means that the probability of one event within a small time interval Δt is approximately $\lambda \Delta t$, where $o(\Delta t)$ is a function of Δt that goes to zero faster than Δt does. That is,

$$\lim_{\Delta t \to 0} \frac{o(\Delta t)}{\Delta t} = 0.$$

2. $P[X(t + \Delta t) - X(t) \geq 2] = o(\Delta t)$, which means that the probability of two or more events within a small time interval Δt is $o(\Delta t)$.
3. $P[X(t + \Delta t) - X(t) = 0] = 1 - \lambda \Delta t + o(\Delta t)$.

These three properties enable us to derive the probability mass function (PMF) of the number of events in a time interval of length t as follows:

$$P[X(t+\Delta t) = n] = P[X(t) = n]P[X(\Delta t) = 0]$$
$$+ P[X(t) = n-1]P[X(\Delta t) = 1] + o(\Delta t)$$
$$= P[X(t) = n](1 - \lambda\Delta t)$$
$$+ P[X(t) = n-1]\lambda\Delta t + o(\Delta t),$$

$$P[X(t+\Delta t) = n] - P[X(t) = n] = -\lambda P[X(t) = n]\Delta t$$
$$+ \lambda P[X(t) = n-1]\Delta t + o(\Delta t),$$

$$\frac{P[X(t+\Delta t) = n] - P[X(t) = n]}{\Delta t} = -\lambda P[X(t) = n]$$
$$+ \lambda P[X(t) = n-1] + \frac{o(\Delta t)}{\Delta t},$$

$$\lim_{\Delta t \to 0}\left\{\frac{P[X(t+\Delta t) = n] - P[X(t) = n]}{\Delta t}\right\} = \frac{d}{dt}P[X(t) = n] = -\lambda P[X(t) = n]$$
$$+ \lambda P[X(t) = n-1],$$

$$\frac{d}{dt}P[X(t) = n] + \lambda P[X(t) = n] = \lambda P[X(t) = n-1].$$

The last equation may be solved iteratively for $n = 0, 1, 2, \ldots$, subject to the initial conditions:

$$P[X(0) = n] = \begin{cases} 1 & n = 0 \\ 0 & n \neq 0. \end{cases}$$

This gives the PMF of the number of events (or "arrivals") in an interval of length t as:

$$p_{X(t)}(n, t) = \frac{(\lambda t)^n}{n!}e^{-\lambda t} \quad t \geq 0, n = 0, 1, 2, \ldots.$$

From the results obtained for Poisson random variables earlier in the chapter, we have that:

$$G_{X(t)}(z) = e^{\lambda t(z-1)},$$
$$E[X(t)] = \lambda t,$$
$$\sigma^2_{X(t)} = \lambda t.$$

The fact that the mean $E[X(t)] = \lambda t$ indicates that λ is the expected number of arrivals per unit time in the Poisson process. Thus, the parameter λ is called the *arrival rate* for the process. If λ is independent of time, the Poisson process is called a *homogeneous Poisson process*. Sometimes the arrival rate is a function of time, and we represent it as $\lambda(t)$. Such processes are called *nonhomogeneous Poisson processes*. In this book we are concerned mainly with homogeneous Poisson processes.

Another important property of the Poisson process is that the interarrival times of customers are exponentially distributed with parameter λ. This can be demonstrated by noting that if T is the time until the next arrival, then $P[T > t] = P[X(t) = 0] = e^{-\lambda t}$. Therefore, the CDF and PDF of the interarrival time are given by:

$$F_T(t) = P[T \leq t] = 1 - P[T > t] = 1 - e^{-\lambda t},$$

$$f_T(t) = \frac{d}{d_t} F_T(t) = \lambda e^{-\lambda t}.$$

Thus, the time until the next arrival is always exponentially distributed. From the memoryless property of the exponential distribution, the future evolution of the Poisson process is independent of the past and is always probabilistically the same. Therefore, the Poisson process is memoryless. As the saying goes, "the Poisson process implies exponential distribution, and the exponential distribution implies the Poisson process." The Poisson process deals with the number of arrivals within a given time interval where interarrival times are exponentially distributed, and the exponential distribution measures the time between arrivals where customers arrive according to a Poisson process.

2.8 RENEWAL PROCESSES

Consider an experiment that involves a set of identical light bulbs whose lifetimes are independent. The experiment consists of using one light bulb at a time, and when it fails it is immediately replaced by another light bulb from the set. Each time a failed light bulb is replaced constitutes a *renewal event*. Let X_i denote the lifetime of the ith light bulb, $i = 1, 2, \ldots$, where $X_0 = 0$. Because the light bulbs are assumed to be identical, the X_i are independent and identically distributed with PDF $f_X(x)$, $x \geq 0$, and mean $E[X]$.

Let $N(t)$ denote the number of renewal events up to and including the time t, where it is assumed that the first light bulb was turned on at time $t = 0$. The time to failure T_n of the first n light bulbs is given by:

$$T_0 = 0,$$
$$T_1 = X_1,$$
$$T_2 = X_1 + X_2,$$
$$\vdots$$
$$T_n = X_1 + \ldots + X_n.$$

The relationship between the interevent times X_n and the T_n is illustrated in Figure 2.2, where E_k denotes the kth event.

T_n is called the time of the nth renewal, and we have that:

$$N(t) = \max\{n | T_n \leq t\}.$$

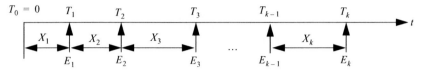

Figure 2.2   Interarrival times of a renewal process.

Thus, the process $\{N(t)|t \geq 0\}$ is a counting process known as a *renewal process*, and $N(t)$ denotes the number of renewals up to time t. Observe that the event that the number of renewals up to and including the time t is less than n is equivalent to the event that the nth renewal occurs at a time that is later than t. Thus, we have that:

$$\{N(t) < n\} = \{T_n > t\}.$$

Therefore, $P[N(t) < n] = P[T_n > t]$. Let $f_{T_n}(t)$ and $F_{T_n}(t)$ denote the PDF and CDF, respectively, of T_n. Thus, we have that:

$$P[N(t) < n] = P[T_n > t] = 1 - F_{T_n}(t).$$

Because $P[N(t) = n] = P[N(t) < n + 1] - P[N(t) < n]$, we obtain the following result for the PMF of $N(t)$:

$$p_{N(t)}(n) = P[N(t) = n] = P[N(t) < n+1] - P[N(t) < n] = 1 - F_{T_{n+1}}(t) - \{1 - F_{T_n}(t)\}$$
$$= F_{T_n}(t) - F_{T_{n+1}}(t).$$

2.8.1   The Renewal Equation

The expected number of renewals by time t is called the *renewal function*. It is denoted by $H(t)$ and given by:

$$H(t) = E[N(t)] = \sum_{n=0}^{\infty} nP[N(t) = n] = \sum_{n=0}^{\infty} n\{F_{T_n}(t) - F_{T_{n+1}}(t)\}$$
$$= \{F_{T_1}(t) + 2F_{T_2}(t) + 3F_{T_3}(t) + ...\} - \{F_{T_2}(t) + 2F_{T_3}(t) + ...\}$$
$$= F_{T_1}(t) + F_{T_2}(t) + F_{T_3}(t) + ...$$
$$= \sum_{n=1}^{\infty} F_{T_n}(t).$$

If we take the derivative of each side we obtain:

$$h(t) = \frac{dH(t)}{dt} = \frac{d}{dt}\sum_{n=1}^{\infty} F_{T_n}(t) = \sum_{n=1}^{\infty}\frac{d}{dt}F_{T_n}(t) = \sum_{n=1}^{\infty} f_{T_n}(t) \quad t \geq 0,$$

where $h(t)$ is called the *renewal density*. Let $M_h(s)$ denote the one-sided Laplace transform of $h(t)$ and $M_{T_n}(s)$ the s-transform of $f_{T_n}(t)$. Because T_n is the sum of n independent and identically distributed random variables, the PDF $f_{T_n}(t)$ is the n-fold convolution of the PDF of X. Thus, we have that $M_{T_n}(s) = \{M_X(s)\}^n$. From this we obtain $M_h(s)$ as follows:

$$M_h(s) = \sum_{n=1}^{\infty} M_{T_n}(s) = \sum_{n=1}^{\infty} \{M_X(s)\}^n$$

$$= \frac{1}{1 - M_X(s)} - 1 = \frac{M_X(s)}{1 - M_X(s)}.$$

This gives:

$$M_h(s) = M_X(s) + M_h(s) M_X(s).$$

Taking the inverse transform we obtain:

$$h(t) = f_X(t) + \int_{u=0}^{t} h(t-u) f_X(u) du.$$

Finally, integrating both sides of the equation, we obtain:

$$H(t) = F_X(t) + \int_{u=0}^{t} H(t-u) f_X(u) du.$$

This equation is called the *fundamental equation of renewal theory*.

Example 2.1: Assume that X is exponentially distributed with mean $1/\lambda$. Then we obtain:

$$f_X(t) = \lambda e^{-\lambda t},$$

$$M_X(s) = \frac{\lambda}{s + \lambda},$$

$$M_h(s) = \frac{M_X(s)}{1 - M_X(s)} = \frac{\lambda}{s},$$

$$h(t) = L^{-1}\{M_h(s)\} = \lambda,$$

$$H(t) = \int_{u=0}^{t} h(u) du = \lambda t,$$

where $L^{-1}\{M_h(s)\}$ is the inverse Laplace transform of $M_h(s)$.

2.8.2   The Elementary Renewal Theorem

We state the following theorem called the elementary renewal theorem without proof:

$$\lim_{t\to\infty}\frac{H(t)}{t}=\frac{1}{E[X]}.$$

2.8.3   Random Incidence and Residual Time

Consider a renewal process $N(t)$ in which events (or arrivals) occur at times $0 = T_0, T_1, T_2, \ldots$. As discussed earlier, the interevent times X_k can be defined in terms of the T_k as follows:

$$X_1 = T_1 - T_0 = T_1,$$
$$X_2 = T_2 - T_1,$$
$$\ldots$$
$$X_k = T_k - T_{k-1}.$$

Note that the X_k are mutually independent and identically distributed.

Consider the following problem in connection with the X_k. Assume the T_k are the points in time that buses arrive at a bus stop. A passenger arrives at the bus stop at a *random time* and wants to know how long he or she will wait until the next bus arrival. This problem is usually referred to as the *random incidence problem*, because the subject (or passenger in this example) is incident to the process at a random time. Let R be the random variable that denotes the time from the moment the passenger arrived until the next bus arrival. R is referred to as the *residual life* of the renewal process. Also, let W denote the length of the interarrival gap that the passenger entered by random incidence. Figure 2.3 illustrates the random incidence problem.

Let $f_X(x)$ denote the PDF of the interarrival times; let $f_W(w)$ denote the PDF of W, the gap entered by random incidence; and let $f_R(r)$ denote the PDF of the residual life, R. The probability that the random arrival occurs in a gap of length between w and $w + dw$ can be assumed to be directly proportional to the length w of the gap and relative occurrence $f_X(w)dw$ of such gaps. That is,

$$f_W(w)dw = \beta w f_X(w)dw,$$

where β is a constant of proportionality. Thus, $f_W(w) = \beta w f_X(w)$. Because $f_W(w)$ is a PDF, we have that:

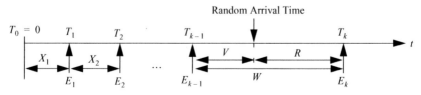

Figure 2.3   Random incidence.

$$\int_{-\infty}^{\infty} f_W(w)\,dw = 1 = \beta \int_{-\infty}^{\infty} w f_X(w)\,dw = \beta E[X].$$

Thus, $\beta = 1/E[X]$, and we obtain:

$$f_W(w) = \frac{w f_X(w)}{E[X]}.$$

The expected value of W is given by $E[W] = E[X^2]/E[X]$. This result applies to all renewal processes.

A Poisson process is an example of a renewal process in which X is exponentially distributed with $E[X] = 1/\lambda$ and $E[X^2] = 2/\lambda^2$. Thus, for a Poisson process we obtain:

$$f_W(w) = \lambda w f_X(w) = \lambda^2 w e^{-\lambda w} \quad w \geq 0,$$
$$E[W] = 2/\lambda.$$

This means that for a Poisson process the gap entered by random incidence has the second-order Erlang distribution; thus, the expected length of the gap is twice the expected length of an interarrival time. This is often referred to as the *random incidence paradox*. The reason for this fact is that the passenger is more likely to enter a large gap than a small gap; that is, the gap entered by random incidence is not a typical interval.

Next, we consider the PDF of the residual life R of the process. Given that the passenger enters a gap of length w, he or she is equally likely to be anywhere within the gap. Thus, the conditional PDF of R, given that $W = w$, is given by:

$$f_{R|W}(r|w) = \frac{1}{w} \quad 0 \leq r \leq w.$$

When we combine this result with the previous one, we get the joint PDF of R and W as follows:

$$f_{RW}(r,w) = f_{R|W}(r|w) f_W(w) = \frac{1}{w} \left\{ \frac{w f_X(w)}{E[X]} \right\}$$
$$= \frac{f_X(w)}{E[X]} \quad 0 \leq r \leq w \leq \infty.$$

The marginal PDF of R and its expected value become:

$$f_R(r) = \int_{-\infty}^{\infty} f_{RW}(r,w)\,dw = \int_{r}^{\infty} \frac{f_X(w)}{E[X]}\,dw = \frac{1 - F_X(r)}{E[X]} \quad r \geq 0,$$

$$E[R] = \int_{0}^{\infty} r f_R(r)\,dr = \frac{1}{E[X]} \int_{r=0}^{\infty} r \int_{w=r}^{\infty} f_X(w)\,dw\,dr = \frac{1}{E[X]} \int_{w=0}^{\infty} \int_{r=0}^{w} r f_X(w)\,dr\,dw$$

$$= \frac{1}{E[X]} \int_{w=0}^{\infty} f_X(w) \left[\frac{r^2}{2} \right]_{0}^{w} dw = \frac{E[X^2]}{2E[X]}.$$

For the Poisson process, X is exponentially distributed and $1 - F_X(r) = e^{-\lambda r}$, which means that:

$$f_R(r) = \lambda e^{-\lambda r} \quad r \geq 0.$$

Thus, for a Poisson process, the residual life of the process has the same distribution as the interarrival time, which can be expected from the "forgetfulness" property of the exponential distribution.

In Figure 2.3, the random variable V denotes the time between the last bus arrival and the passenger's random arrival. Because $W = V + R$, the expected value of V is:

$$E[V] = E[W] - E[R] = \frac{E[X^2]}{E[X]} - \frac{E[X^2]}{2E[X]} = \frac{E[X^2]}{2E[X]} = E[R].$$

2.9   MARKOV PROCESSES

Markov processes are widely used in engineering, science, and business modeling. They are used to model systems that have a limited memory of their past. For example, consider a sequence of games where a player gets \$1 if he wins a game and loses \$1 if he loses the game. Then the amount of money the player will make after $n + 1$ games is determined by the amount of money he has made after n games. Any other information is irrelevant in making this prediction. In population growth studies, the population of the next generation depends mainly on the current population and possibly the last few generations.

A random process $\{X(t) | t \in T\}$ is called a first-order Markov process if for any $t_0 < t_1 < \ldots < t_n$ the conditional CDF of $X(t_n)$ for given values of $X(t_0)$, $X(t_1), \ldots, X(t_{n-1})$ depends only on $X(t_{n-1})$. That is,

$$P[X(t_n) \leq x_n | X(t_{n-1}) = x_{n-1}, X(t_{n-2}) = x_{n-2}, \ldots, X(t_0) = x_0]$$
$$= P[X(t_n) \leq x_n | X(t_{n-1}) = x_{n-1}].$$

This means that, given the present state of the process, the future state is independent of the past. This property is usually referred to as the *Markov property*. In second-order Markov processes, the future state depends on both the current state and the last immediate state, and so on for higher order Markov processes. In this chapter we consider only first-order Markov processes.

Markov processes are classified according to the nature of the time parameter and the nature of the state space. With respect to state space, a Markov process can be either a discrete-state Markov process or a continuous-state Markov process. A discrete-state Markov process is called a *Markov chain*. Similarly, with respect to time, a Markov process can be either a discrete-time

		State Space	
		Discrete	Continuous
Time	Discrete	Discrete-Time Markov Chain	Discrete-Time Markov Process
	Continuous	Continuous-Time Markov Chain	Continuous-Time Markov Process

Figure 2.4   Classification of Markov processes.

Markov process or a continuous-time Markov process. Thus, there are four basic types of Markov processes (Ibe 2009):

1. Discrete-time Markov chain (or discrete-time discrete-state Markov process)
2. Continuous-time Markov chain (or continuous-time discrete-state Markov process)
3. Discrete-time Markov process (or discrete-time continuous-state Markov process)
4. Continuous-time Markov process (or continuous-time continuous-state Markov process)

This classification of Markov processes is illustrated in Figure 2.4.

2.9.1   Discrete-Time Markov Chains

The discrete-time process $\{X_k, k = 0, 1, 2, \ldots \}$ is called a Markov chain if for all $i, j, k, \ldots, m$, the following is true:

$$P[X_k = j | X_{k-1} = i, X_{k-2} = n, \ldots, X_0 = m] = P[X_k = j | X_{k-1} = i] = p_{ijk}.$$

The quantity p_{ijk} is called the *state transition probability*, which is the conditional probability that the process will be in state j at time k immediately after the next transition, given that it is in state i at time $k - 1$. A Markov chain that obeys the preceding rule is called a *nonhomogeneous Markov chain*. In this book we will consider only *homogeneous Markov chains*, which are Markov chains in which $p_{ijk} = p_{ij}$. This means that homogeneous Markov chains do not depend on the time unit, which implies that:

$$P[X_k = j | X_{k-1} = i, X_{k-2} = \alpha, \ldots, X_0 = \theta] = P[X_k = j | X_{k-1} = i] = p_{ij},$$

which is the so-called *Markov property*. The *homogeneous state transition probability* p_{ij} satisfies the following conditions:

1. $0 \leq p_{ij} \leq 1$
2. $\sum_j p_{ij} = 1, i = 1, 2, \ldots, n$, which follows from the fact that the states are mutually exclusive and collectively exhaustive.

From the above definition we obtain the following *Markov chain rule*:

$$
\begin{aligned}
P[X_k &= j, X_{k-1} = i_1, X_{k-2} \ldots, X_0] \\
&= P[X_k = j | X_{k-1} = i_1, X_{k-2} \ldots, X_0] P[X_{k-1} = i_1, X_{k-2} = i_2 \ldots, X_0 = i_k] \\
&= P[X_k = j | X_{k-1} = i_1] P[X_{k-1} = i_1, X_{k-2} \ldots, X_0 = i_k] \\
&= P[X_k = j | X_{k-1} = i_1] P[X_{k-1} = i_1 | X_{k-2} = i_2 \ldots, X_0] P[X_{k-2} = i_2 \ldots, X_0] \\
&= P[X_k = j | X_{k-1} = i_1] P[X_{k-1} = i_1 | X_{k-2} = i_2] \ldots P[X_1 = i_{k-1} | X_0] P[X_0 = i_k] \\
&= p_{i_1 j} p_{i_2 i_1} p_{i_3 i_2} \ldots p_{i_k i_{k-1}} P[X_0 = i_k].
\end{aligned}
$$

Thus, once we know the initial state X_0 we can evaluate the joint probability $P[X_k, X_{k-1}, \ldots, X_0]$.

2.9.1.1 State Transition Probability Matrix

It is customary to display the state transition probabilities as the entries of an $n \times n$ matrix P, where p_{ij} is the entry in the ith row and jth column:

$$
P = \begin{bmatrix}
p_{11} & p_{12} & \cdots & p_{1n} \\
p_{21} & p_{22} & \cdots & p_{2n} \\
\cdots & \cdots & \cdots & \cdots \\
p_{n1} & p_{n2} & \cdots & p_{nn}
\end{bmatrix}.
$$

P is called the transition probability matrix. It is a stochastic matrix because for any row i, $\sum_i p_{ij} = 1$.

2.9.1.2 The n-Step State Transition Probability

Let $p_{ij}(n)$ denote the conditional probability that the system will be in state j after exactly n transitions, given that it is currently in state i. That is,

$$
\begin{aligned}
p_{ij}(n) &= P[X_{m+n} = j | X_m = i], \\
p_{ij}(0) &= \begin{cases} 1 & i = j \\ 0 & i \neq j \end{cases}, \\
p_{ij}(1) &= p_{ij}.
\end{aligned}
$$

Consider the two-step transition probability $p_{ij}(2)$, which is defined by:

$$
p_{ij}(2) = P[X_{m+2} = j | X_m = i].
$$

Assume that $m = 0$, then:

$$p_{ij}(2) = P[X_2 = j | X_0 = i] = \sum_k P[X_2 = j, X_1 = k | X_0 = i]$$

$$= \sum_k P[X_2 = j | X_1 = k, X_0 = i] P[X_1 = k | X_0 = i]$$

$$= \sum_k P[X_2 = j | X_1 = k] P[X_1 = k | X_0 = i]$$

$$= \sum_k p_{kj} p_{ik} = \sum_k p_{ik} p_{kj},$$

where the second to the last equality is due to the Markov property. The final equation states that the probability of starting in state i and being in state j at the end of the second transition is the probability that we first go immediately from state i to some intermediate state k and then immediately from state k to state j; the summation is taken over all possible intermediate states k.

Proposition: The following proposition deals with a class of equations called the *Chapman–Kolmogorov equations*, which provide a generalization of the above results obtained for the two-step transition probability. For all $0 < r < n$,

$$p_{ij}(n) = \sum_k p_{ik}(r) p_{kj}(n-r).$$

This proposition states that the probability that the process starts in state i and finds itself in state j at the end of the nth transition is the product of the probability that the process starts in state i and finds itself in some intermediate state k after r transitions and the probability that it goes from state k to state j after additional $n - r$ transitions.

Proof: The proof is a generalization of the proof for the case of $n = 2$ and is as follows:

$$p_{ij}(n) = P[X_n = j | X_0 = i] = \sum_k P[X_n = j, X_r = k | X_0 = i]$$

$$= \sum_k P[X_n = j | X_r = k, X_0 = i] P[X_r = k | X_0 = i]$$

$$= \sum_k P[X_n = j | X_r = k] P[X_r = k | X_0 = i] = \sum_k p_{kj}(n-r) p_{ik}(r)$$

$$= \sum_k p_{ik}(r) p_{kj}(n-r).$$

From the preceding discussion it can be shown that $p_{ij}(n)$ is the ijth entry (ith row, jth column) in the matrix P^n. That is, for an N-state Markov chain, P^n is the matrix:

$$P^n = \begin{bmatrix} p_{11}(n) & p_{12}(n) & p_{13}(n) & \cdots & p_{1N}(n) \\ p_{21}(n) & p_{22}(n) & p_{23}(n) & \cdots & p_{2N}(n) \\ p_{31}(n) & p_{32}(n) & p_{33}(n) & \cdots & p_{3N}(n) \\ \cdots & \cdots & \cdots & \cdots & \cdots \\ p_{N1}(n) & p_{N2}(n) & p_{N3}(n) & \cdots & p_{NN}(n) \end{bmatrix}.$$

2.9.1.3 State Transition Diagrams Consider the following problem. It has been observed via a series of tosses of a particular biased coin that the outcome of the next toss depends on the outcome of the current toss. In particular, given that the current toss comes up heads, the next toss will come up heads with probability 0.6 and tails with probability 0.4. Similarly, given that the current toss comes up tails, the next toss will come up heads with probability 0.35 and tails with probability 0.65.

If we define state 1 to represent heads and state 2 to represent tails, then the transition probability matrix for this problem is the following:

$$P = \begin{bmatrix} 0.6 & 0.4 \\ 0.35 & 0.65 \end{bmatrix}.$$

All the properties of the Markov process can be determined from this matrix. However, the analysis of the problem can be simplified by the use of the *state transition diagram* in which the states are represented by circles and directed arcs represent transitions between states. The state transition probabilities are labeled on the appropriate arcs. Thus, with respect to the above problem, we obtain the state transition diagram shown in Figure 2.5.

Example 2.2: Assume that people in a particular society can be classified as belonging to the upper class (U), middle class (M), and lower class (L). Membership in any class is inherited in the following probabilistic manner. Given that a person is raised in an upper-class family, he or she will have an upper-class family with probability 0.7, a middle-class family with probability 0.2, and a lower-class family with probability 0.1. Similarly, given that a person is raised in a middle-class family, he or she will have an upper-class family with probability 0.1, a middle-class family with probability 0.6, and a lower-class family with probability 0.3. Finally, given that a person is raised in a lower-class family, he or she will have a middle-class family with probability 0.3 and a

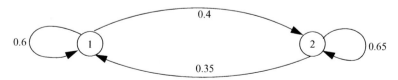

Figure 2.5 Example of a state transition diagram.

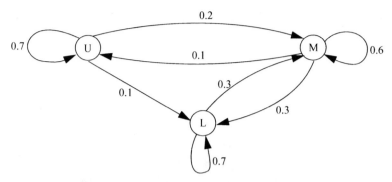

Figure 2.6 State transition diagram for Example 2.2.

lower-class family with probability 0.7. Determine (a) the transition probability matrix and (b) the state transition diagram for this problem.

Solution: (a) Using the first row to represent the upper class, the second row to represent the middle class, and the third row to represent the lower class, we obtain the following transition probability matrix:

$$P = \begin{bmatrix} 0.7 & 0.2 & 0.1 \\ 0.1 & 0.6 & 0.3 \\ 0.0 & 0.3 & 0.7 \end{bmatrix}.$$

(b) The state transition diagram is as shown in Figure 2.6.

2.9.1.4 Classification of States A state j is said to be *accessible* (or *can be reached*) from state i if, starting from state i, it is possible that the process will ever enter state j. This implies that $p_{ij}(n) > 0$ for some $n > 0$. Thus, the n-step probability enables us to obtain reachability information between any two states of the process.

Two states that are accessible from each other are said to *communicate* with each other. The concept of communication divides the state space into different classes. Two states that communicate are said to be in the same *class*. All members of one class communicate with one another. If a class is not accessible from any state outside the class, we define the class to be a *closed communicating class*. A Markov chain in which all states communicate, which means that there is only one class, is called an *irreducible* Markov chain. For example, the Markov chains shown in Figures 2.5 and 2.6 are irreducible Markov chains.

The states of a Markov chain can be classified into two broad groups: those that the process enters infinitely often and those that it enters finitely often. In the long run, the process will be found to be in only those states that it enters infinitely often. Let $f_{ij}(n)$ denote the conditional probability that given that the process is presently in state i, the first time it will enter state j occurs in exactly n transitions (or steps). We call $f_{ij}(n)$ the probability of *first passage* from state i to state j in n transitions. The parameter f_{ij}, which is defined as follows,

$$f_{ij} = \sum_{n=1}^{\infty} f_{ij}(n),$$

is the probability of first passage from state i to state j. It is the conditional probability that the process will ever enter state j, given that it was initially in state i. Obviously $f_{ij}(1) = p_{ij}$ and a recursive method of computing $f_{ij}(n)$ is:

$$f_{ij}(n) = \sum_{l \neq j} p_{il} f_{lj}(n-1).$$

The quantity f_{ii} denotes the probability that a process that starts at state i will ever return to state i. Any state i for which $f_{ii} = 1$ is called a *recurrent state*, and any state i for which $f_{ii} < 1$ is called a *transient state*. More formally, we define these states as follows:

a. A state j is called a *transient* (or *nonrecurrent*) state if there is a positive probability that the process will never return to j again after it leaves j.

b. A state j is called a *recurrent* (or *persistent*) state if, with probability 1, the process will eventually return to j after it leaves j. A set of recurrent states forms a *single chain* if every member of the set communicates with all other members of the set.

c. A recurrent state j is called a *periodic* state if there exists an integer d, $d > 1$, such that $p_{jj}(n)$ is zero for all values of n other than $d, 2d, 3d, \ldots$; d is called the period. If $d = 1$, the recurrent state j is said to be *aperiodic*.

d. A recurrent state j is called a *positive recurrent* state if, starting at state j, the expected time until the process returns to state j is finite. Otherwise, the recurrent state is called a *null recurrent* state.

e. Positive recurrent, aperiodic states are called *ergodic* states.

f. A chain consisting of ergodic states is called an *ergodic chain*.

g. A state j is called an *absorbing* (or *trapping*) state if $p_{jj} = 1$. Thus, once the process enters a trapping or absorbing state, it never leaves the state, which means that it is "trapped."

Example 2.3: Consider the Markov chain with the state transition diagram shown in Figure 2.7. State 4 is a transient state while states 1, 2, and 3 are recurrent states. There is no periodic state, and there is one chain, which is $\{1, 2, 3\}$.

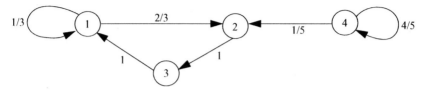

Figure 2.7   State transition diagram for Example 2.3.

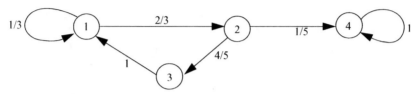

Figure 2.8   State transition diagram for Example 2.4.

Example 2.4: Consider the state transition diagram of Figure 2.8, which is a modified version of Figure 2.7. Here, the transition is now from state 2 to state 4 instead of from state 4 to state 2. For this case, states 1, 2, and 3 are now transient states because when the process enters state 2 and makes a transition to state 4, it does not return to these states again. Also, state 4 is a trapping (or absorbing) state because once the process enters the state, the process never leaves the state. As stated in the definition, we identify a trapping state by the fact that, as in this example, $p_{44} = 1$ and $p_{4k} = 0$ for k not equal to 4.

2.9.1.5   Limiting State Probabilities   Recall that the n-step state transition probability $p_{ij}(n)$ is the conditional probability that the system will be in state j after exactly n transitions, given that it is presently in state i. The n-step transition probabilities can be obtained by multiplying the transition probability matrix by itself n times. For example, consider the following transition probability matrix:

$$P = \begin{bmatrix} 0.4 & 0.5 & 0.1 \\ 0.3 & 0.3 & 0.4 \\ 0.3 & 0.2 & 0.5 \end{bmatrix},$$

$$P^2 = \begin{bmatrix} 0.4 & 0.5 & 0.1 \\ 0.3 & 0.3 & 0.4 \\ 0.3 & 0.2 & 0.5 \end{bmatrix} \times \begin{bmatrix} 0.4 & 0.5 & 0.1 \\ 0.3 & 0.3 & 0.4 \\ 0.3 & 0.2 & 0.5 \end{bmatrix} = \begin{bmatrix} 0.34 & 0.37 & 0.29 \\ 0.33 & 0.32 & 0.35 \\ 0.33 & 0.31 & 0.36 \end{bmatrix},$$

$$P^3 = \begin{bmatrix} 0.34 & 0.37 & 0.29 \\ 0.33 & 0.32 & 0.35 \\ 0.33 & 0.31 & 0.36 \end{bmatrix} \times \begin{bmatrix} 0.4 & 0.5 & 0.1 \\ 0.3 & 0.3 & 0.4 \\ 0.3 & 0.2 & 0.5 \end{bmatrix} = \begin{bmatrix} 0.334 & 0.339 & 0.327 \\ 0.333 & 0.331 & 0.336 \\ 0.333 & 0.330 & 0.337 \end{bmatrix}.$$

From the matrix P^2 we obtain the $p_{ij}(2)$. For example, $p_{23}(2) = 0.35$, which is the entry in the second row and third column of the matrix P^2. Similarly, the entries of the matrix P^3 are the $p_{ij}(3)$.

For this particular matrix and matrices for a large number of Markov chains, we find that as we multiply the transition probability matrix by itself many times, the entries remain constant. More importantly, all the members of one column will tend to converge to the same value.

If we define $P[X(0) = i]$ as the probability that the process is in state i before it makes the first transition, then the set $\{P[X(0) = i]\}$ defines the initial condition for the process, and for an N-state process,

$$\sum_{i=1}^{N} P[X(0) = i] = 1.$$

Let $P[X(n) = j]$ denote the probability that the process is in state j at the end of the first n transitions, then for the N-state process,

$$P[X(n) = j] = \sum_{i=1}^{N} P[X(0) = i] p_{ij}(n).$$

For the class of Markov chains referenced above, it can be shown that as $n \rightarrow \infty$, the n-step transition probability $p_{ij}(n)$ does not depend on i, which means that $P[X(n) = j]$ approaches a constant as $n \rightarrow \infty$ for this class of Markov chains. That is, the constant is independent of the initial conditions. Thus, for the class of Markov chains in which the limit exists, we define the *limiting state probabilities* as follows:

$$\lim_{n \rightarrow \infty} P[X(n) = j] = \pi_j \quad j = 1, 2, \ldots, N.$$

Recall that the n-step transition probability can be written in the form:

$$p_{ij}(n) = \sum_{k} p_{ik}(n-1) p_{kj}.$$

If the limiting state probabilities exist and do not depend on the initial state, then:

$$\lim_{n \rightarrow \infty} p_{ij}(n) = \pi_j = \lim_{n \rightarrow \infty} \sum_{k} p_{ik}(n-1) p_{kj} = \sum_{k} \pi_k p_{kj}.$$

If we define the limiting state probability vector $\pi = [\pi_1, \pi_2, \ldots, \pi_N]$, then we have that:

$$\pi_j = \sum_{k} \pi_k p_{kj},$$

$$\pi = \pi P,$$

$$1 = \sum_{j} \pi_j,$$

where the last equation is due to the law of total probability. Each of the first two equations, together with the last equation, gives a system of linear equations that the π_j must satisfy. The following propositions specify the conditions for the existence of the limiting state probabilities:

a. In any irreducible, aperiodic Markov chain, the limits $\pi_j = \lim_{n \to \infty} p_{ij}(n)$ exist and are independent of the initial distribution.

b. In any irreducible, periodic Markov chain the limits $\pi_j = \lim_{n \to \infty} p_{ij}(n)$ exist and are independent of the initial distribution. However, they must be interpreted as the long-run probability that the process is in state j.

Example 2.5: Recall the biased coin problem whose state transition diagram is given in Figure 2.5 and reproduced in Figure 2.9. Suppose we are required to find the limiting state probabilities. We proceed as follows. There are three equations associated with the above Markov chain, and they are:

$$\pi_1 = 0.6\pi_1 + 0.35\pi_2,$$
$$\pi_2 = 0.4\pi_1 + 0.65\pi_2,$$
$$1 = \pi_1 + \pi_2.$$

Since there are three equations and two unknowns, one of the equations is redundant. Thus, the rule of thumb is that for an N-state Markov chain, we use the first $N - 1$ linear equations from the relation $\pi_j = \sum_k \pi_k p_{kj}$ and the law of total probability: $1 = \sum_j \pi_j$. For the given problem we have:

$$\pi_1 = 0.6\pi_1 + 0.35\pi_2,$$
$$1 = \pi_1 + \pi_2.$$

From the first equation, we obtain $\pi_1 = (0.35/0.4)\pi_2 = (7/8)\pi_2$. Substituting for π_1 and solving for π_2 in the second equation, we obtain the result $\pi = \{\pi_1, \pi_2\} = \{7/15, 8/15\}$.

Suppose we are also required to compute $p_{12}(3)$, which is the probability that the process will be in state 2 at the end of the third transition, given that it is presently in state 1. We can proceed in two ways: the direct method and the matrix method. We consider both methods.

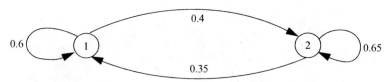

Figure 2.9   State transition diagram for Example 2.5.

a. *Direct Method:* Under this method we exhaustively enumerate all the possible ways of a state 1-to-state 2 transition in three steps. If we use the notation $a \rightarrow b \rightarrow c$ to denote a transition from state a to state b and then from state b to state c, the desired result is the following:

$$p_{12}(3) = P[\{1 \rightarrow 1 \rightarrow 1 \rightarrow 2\} \cup \{1 \rightarrow 1 \rightarrow 2 \rightarrow 2\} \cup \{1 \rightarrow 2 \rightarrow 1 \rightarrow 2\}$$
$$\cup \{1 \rightarrow 2 \rightarrow 2 \rightarrow 2\}].$$

Since the different events are mutually exclusive, we obtain:

$$p_{12}(3) = P[1 \rightarrow 1 \rightarrow 1 \rightarrow 2] + P[1 \rightarrow 1 \rightarrow 2 \rightarrow 2] + P[1 \rightarrow 2 \rightarrow 1 \rightarrow 2]$$
$$+ P[1 \rightarrow 2 \rightarrow 2 \rightarrow 2]$$
$$= (0.6)(0.6)(0.4) + (0.6)(0.4)(0.65) + (0.4)(0.35)(0.4) + (0.4)(0.65)(0.65)$$
$$= 0.525.$$

b. *Matrix Method:* One of the limitations of the direct method is that it is difficult to exhaustively enumerate the different ways of going from state 1 to state 2 in n steps, especially when n is large. This is where the matrix method becomes very useful. As discussed earlier, $p_{ij}(n)$ is the ijth entry in the matrix P^n. Thus, for the current problem, we are looking for the entry in the first row and second column of the matrix P^3. Therefore, we have:

$$P = \begin{bmatrix} 0.6 & 0.4 \\ 0.35 & 0.65 \end{bmatrix},$$

$$P^2 = P \times P = \begin{bmatrix} 0.6 & 0.4 \\ 0.35 & 0.65 \end{bmatrix} \times \begin{bmatrix} 0.6 & 0.4 \\ 0.35 & 0.65 \end{bmatrix} = \begin{bmatrix} 0.5 & 0.5 \\ 0.4375 & 0.5625 \end{bmatrix},$$

$$P^3 = P \times P^2 = \begin{bmatrix} 0.6 & 0.4 \\ 0.35 & 0.65 \end{bmatrix} \times \begin{bmatrix} 0.5 & 0.5 \\ 0.4375 & 0.5625 \end{bmatrix} = \begin{bmatrix} 0.475 & 0.525 \\ 0.459375 & 0.540625 \end{bmatrix}.$$

The required result (first row, second column) is 0.525, which is the result obtained via the direct method.

2.9.1.6 Doubly Stochastic Matrix
A transition probability matrix P is defined to be a doubly stochastic matrix if each of its columns sums to 1. That is, not only does each row sum to 1, each column also sums to 1. Thus, for every column j of a doubly stochastic matrix, we have that $\sum_i p_{ij} = 1$.

Doubly stochastic matrices have interesting limiting state probabilities, as the following theorem shows.

Theorem: If P is a doubly stochastic matrix associated with the transition probabilities of a Markov chain with N states, then the limiting state probabilities are given by $\pi_i = 1/N$, $i = 1, 2, \ldots, N$.

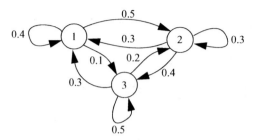

Figure 2.10 State transition diagram for Example 2.6.

Proof: We know that the limiting state probabilities satisfy the condition:

$$\pi_j = \sum_k \pi_k p_{kj}.$$

To check the validity of the theorem, we observe that when we substitute $\pi_i = 1/N, i = 1, 2, \ldots, N$, in the above equation we obtain:

$$\frac{1}{N} = \frac{1}{N} \sum_k p_{kj}.$$

This shows that $\pi_i = 1/N$ satisfies the condition $\pi = \pi P$, which the limiting state probabilities are required to satisfy. Conversely, from the above equation, we see that if the limiting state probabilities are given by $1/N$, then each column j of P sums to 1; that is, P is doubly stochastic. This completes the proof.

Example 2.6: Find the transition probability matrix and the limiting state probabilities of the process represented by the state transition diagram shown in Figure 2.10.

Solution: The transition probability matrix is given by:

$$P = \begin{bmatrix} 0.4 & 0.5 & 0.1 \\ 0.3 & 0.3 & 0.4 \\ 0.3 & 0.2 & 0.5 \end{bmatrix}.$$

It can be seen that each row of the matrix sums to 1 and each column also sums to 1; that is, it is a doubly stochastic matrix. Since the process is an irreducible, aperiodic Markov chain, the limiting state probabilities exist and are given by $\pi_1 = \pi_2 = \pi_3 = 1/3$.

2.9.2 Continuous-Time Markov Chains

A random process $\{X(t)|t \geq 0\}$ is a continuous-time Markov chain if, for all s, $t \geq 0$ and nonnegative integers i, j, k,

$$P[X(t+s)= j| X(s)=i, X(u)=k, 0 \le u \le s] = P[X(t+s)= j| X(s)=i].$$

This means that in a continuous-time Markov chain, the conditional probability of the future state at time $t + s$ given the present state at s and all past states depends only on the present state and is independent of the past. If, in addition $P[X(t + s) = j|X(s) = i]$ is independent of s, then the process $\{X(t)|t \ge 0\}$ is said to be *time homogeneous* or have the *time homogeneity property*. Time-homogeneous Markov chains have stationary (or homogeneous) transition probabilities. Let:

$$p_{ij}(t) = P[X(t+s)= j| X(s)=i],$$
$$p_j(t) = P[X(t)= j].$$

That is, $p_{ij}(t)$ is the probability that a Markov chain that is presently in state i will be in state j after an additional time t, and $p_j(t)$ is the probability that a Markov chain is in state j at time t. Thus, the $p_{ij}(t)$ are the *transition probability functions* that satisfy the following condition:

$$\sum_j p_{ij}(t) = 1, \quad 0 \le p_{ij}(t) \le 1.$$

Also,

$$\sum_j p_j(t) = 1,$$

which follows from the fact that at any given time the process must be in some state. Also,

$$p_{ij}(t+s) = \sum_k P[X(t+s)= j, X(t)=k| X(0)=i]$$

$$= \sum_k \left\{ \frac{P[X(0)=i, X(t)=k, X(t+s)= j]}{P[X(0)=i]} \right\}$$

$$= \sum_k \left\{ \frac{P[X(0)=i, X(t)=k]}{P[X(0)=i]} \right\} \left\{ \frac{P[X(0)=i, X(t)=k, X(t+s)= j]}{P[X(0)=i, X(t)=k]} \right\}$$

$$= \sum_k P[X(t)=k| X(0)=i] P[X(t+s)= j| X(0)=i, X(t)=k]$$

$$= \sum_k P[X(t)=k| X(0)=i] P[X(t+s)= j| X(t)=k]$$

$$= \sum_k p_{ik}(t) p_{kj}(s).$$

This equation is called the Chapman–Kolmogorov equation for the continuous-time Markov chain. Note that the second to last equation is due to the Markov property. If we define $P(t)$ as the matrix of the $p_{ij}(t)$, that is,

$$P(t) = \begin{bmatrix} p_{11}(t) & p_{12}(t) & p_{13}(t) & \cdots \\ p_{21}(t) & p_{22}(t) & p_{23}(t) & \cdots \\ p_{31}(t) & p_{32}(t) & p_{33}(t) & \cdots \\ \cdots & \cdots & \cdots & \cdots \end{bmatrix},$$

then the Chapman–Kolmogorov equation becomes:

$$P(t+s) = P(t)P(s).$$

Whenever a continuous-time Markov chain enters a state i, it spends an amount of time called the *dwell time* (or *holding time*) in that state. The holding time in state i is exponentially distributed with mean $1/v_i$. At the expiration of the holding time, the process makes a transition to another state j with probability p_{ij}, where:

$$\sum_j p_{ij} = 1.$$

Because the mean holding time in state i is $1/v_i$, v_i represents the rate at which the process leaves state i and $v_i p_{ij}$ represents the rate when in state i that the process makes a transition to state j. Also, since the holding times are exponentially distributed, the probability that when the process is in state i a transition to state $j \neq i$ will take place in the next small time Δt is $p_{ij} v_i \Delta t$. The probability that no transition out of state i will take place in Δt given that the process is presently in state i is $1 - \sum_{j \neq i} p_{ij} v_i \Delta t$, and $\sum_{i \neq i} p_{ij} v_i \Delta t$ is the probability that it leaves state i in Δt.

With these definitions, we consider the state transition diagram for the process, which is shown in Figure 2.11 for state i. We consider the transition equations for state i for the small time interval Δt.

From Figure 2.11, we obtain the following equation:

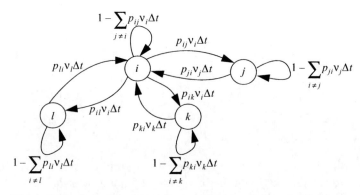

Figure 2.11 State transition diagram for state i over small time Δt.

$$p_i(t+\Delta t) = p_i(t)\left\{1 - \sum_{j\neq i} p_{ij} v_i \Delta t\right\} + \sum_{j\neq i} p_j(t) p_{ji} v_j \Delta t$$

$$p_i(t+\Delta t) - p_i(t) = -p_i(t)\sum_{j\neq i} p_{ij} v_i \Delta t + \sum_{j\neq i} p_j(t) p_{ji} v_j \Delta t$$

$$\frac{p_i(t+\Delta t) - p_i(t)}{\Delta t} = -v_i p_i(t)\sum_{j\neq i} p_{ij} + \sum_{j\neq i} p_j(t) p_{ji} v_j$$

$$\lim_{\Delta t \to 0}\left\{\frac{p_i(t+\Delta t) - p_i(t)}{\Delta t}\right\} = \frac{dp_i(t)}{dt} = -v_i p_i(t)\sum_{j\neq i} p_{ij} + \sum_{j\neq i} p_j(t) p_{ji} v_j.$$

In the steady state, $p_j(t) \to p_j$ and:

$$\lim_{t\to\infty}\left\{\frac{dp_i(t)}{dt}\right\} = 0.$$

Thus, we obtain:

$$0 = -v_i p_i \sum_{j\neq i} p_{ij} + \sum_{j\neq i} p_j p_{ji} v_j,$$

$$1 = \sum_i p_i.$$

Alternatively, we may write:

$$v_i p_i \sum_{j\neq i} p_{ij} = \sum_{j\neq i} p_j p_{ji} v_j,$$

$$1 = \sum_i p_i.$$

The left side of the first equation is the rate of transition out of state i, while the right side is the rate of transition into state i. This "balance" equation states that in the steady state the two rates are equal for any state in the Markov chain.

2.9.2.1 Birth and Death Processes Birth and death processes are a special type of continuous-time Markov chains. Consider a continuous-time Markov chain with states $0, 1, 2, \ldots$. If $p_{ij} = 0$ whenever $j \neq i - 1$ or $j \neq i + 1$, then the Markov chain is called a birth and death process. Thus, a birth and death process is a continuous-time Markov chain with states $0, 1, 2, \ldots$, in which transitions from state i can only go to either state $i + 1$ or state $i - 1$. That is, a transition either causes an increase in state by 1 or a decrease in state by 1. A birth is said to occur when the state increases by 1, and a death is said to occur when the state decreases by 1. For a birth and death process, we define the following *transition rates* from state i:

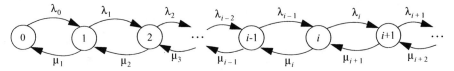

Figure 2.12   State transition rate diagram for the birth and death process.

$$\lambda_i = v_i p_{i(i+1)},$$
$$\mu_i = v_i p_{i(i-1)}.$$

Thus, λ_i is the rate at which a birth occurs when the process is in state i and μ_i is the rate at which a death occurs when the process is in state i. The sum of these two rates is $\lambda_i + \mu_i = v_i$, which is the rate of transition out of state i. The *state transition rate diagram* of a birth and death process is shown in Figure 2.12. It is called a state transition rate diagram as opposed to a state transition diagram because it shows the rate at which the process moves from state to state and not the probability of moving from one state to another. Note that $\mu_0 = 0$, since there can be no death when the process is in an empty state.

The actual state transition probabilities when the process is in state i are $p_{i(i+1)}$ and $p_{i(i-1)}$. By definition, $p_{i(i+1)} = \lambda_i/(\lambda_i + \mu_i)$ is the probability that a birth occurs before a death when the process is in state i. Similarly, $p_{i(i-1)} = \mu_i/(\lambda_i + \mu_i)$ is the probability that a death occurs before a birth when the process is in state i.

Recall that the rate at which the probability of the process being in state i changes with time is given by:

$$\frac{dp_i(t)}{dt} = -v_i p_i(t) \sum_{j \neq i} p_{ij} + \sum_{j \neq i} p_j(t) p_{ji} v_j$$
$$= -(\lambda_i + \mu_i) p_i(t) + \mu_{i+1} p_{i+1}(t) + \lambda_{i-1} p_{i-1}(t).$$

Thus, for the birth and death process we have that:

$$\frac{dp_0(t)}{dt} = -\lambda_0 p_0(t) + \mu_1 p_1(t)$$
$$\frac{dp_i(t)}{dt} = -(\lambda_i + \mu_i) p_i(t) + \mu_{i+1} p_{i+1}(t) + \lambda_{i-1} p_{i-1}(t), \quad i > 0.$$

In the steady state,

$$\lim_{t \to \infty} \left\{ \frac{dp_i(t)}{dt} \right\} = 0.$$

If we assume that the limiting probabilities $\lim_{t \to \infty} p_{ij}(t) = p_j$ exist, then from the above equation we obtain the following:

$$\lambda_0 p_0 = \mu_1 p_1,$$
$$(\lambda_i + \mu_i) = \mu_{i+1} p_{i+1} + \lambda_{i-1} p_{i-1}, i = 1, 2, \dots,$$
$$\sum_i p_i = 1.$$

The equation states that the rate at which the process leaves state i either through a birth or a death is equal to the rate at which it enters the state through a birth when the process is in state $i - 1$ or through a death when the process is in state $i + 1$. This is called the *balance equation* because it balances (or equates) the rate at which the process enters state i with the rate at which it leaves state i.

Example 2.7: A machine is operational for an exponentially distributed time with mean $1/\lambda$ before breaking down. When it breaks down, it takes a time that is exponentially distributed with mean $1/\mu$ to repair it. What is the fraction of time that the machine is operational (or available)?

Solution: This is a two-state birth and death process. Let U denote the up state and D the down state. Then, the state transition rate diagram is shown in Figure 2.13.

Let p_U denote the steady-state probability that the process is in the operational state, and let p_D denote the steady-state probability that the process is in the down state. Then the balance equations become:

$$\lambda p_U = \mu p_D,$$
$$p_U + p_D = 1 \Rightarrow p_D = 1 - p_U.$$

Substituting $p_D = 1 - p_U$ in the first equation gives $p_U = \mu/(\lambda + \mu)$.

Example 2.8: Customers arrive at a bank according to a Poisson process with rate λ. The time to serve each customer is exponentially distributed with mean $1/\mu$. There is only one teller at the bank, and an arriving customer who finds the teller busy when she arrives will join a single queue that operates on a first-come, first-served basis. Determine the limiting state probabilities given that $\mu > \lambda$.

Solution: This is a continuous-time Markov chain in which arrivals constitute births and service completions constitute deaths. Also, for all i, $\mu_i = \mu$ and

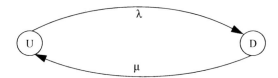

Figure 2.13 State transition rate diagram for Example 2.7.

$\lambda_i = \lambda$. Thus, if p_k denotes the steady-state probability that there are k customers in the system, the balance equations are as follows:

$$\lambda p_0 = \mu p_1 \Rightarrow p_1 = \left(\frac{\lambda}{\mu}\right) p_0,$$

$$(\lambda + \mu)\, p_1 = \lambda p_0 + \mu p_2 \Rightarrow p_2 = \left(\frac{\lambda}{\mu}\right) p_1 = \left(\frac{\lambda}{\mu}\right)^2 p_0,$$

$$(\lambda + \mu)\, p_2 = \lambda p_1 + \mu p_3 \Rightarrow p_3 = \left(\frac{\lambda}{\mu}\right) p_2 = \left(\frac{\lambda}{\mu}\right)^3 p_0.$$

Similarly, it can be shown that:

$$p_k = \left(\frac{\lambda}{\mu}\right)^k p_0 \quad k = 0, 1, 2, \ldots.$$

Now,

$$\sum_{k=0}^{\infty} p_k = 1 = p_0 \sum_{k=0}^{\infty} \left(\frac{\lambda}{\mu}\right)^k = \frac{p_0}{1 - \dfrac{\lambda}{\mu}}.$$

Thus,

$$p_0 = 1 - \frac{\lambda}{\mu},$$

$$p_k = \left(1 - \frac{\lambda}{\mu}\right)\left(\frac{\lambda}{\mu}\right)^k \quad k = 0, 1, 2, \ldots.$$

2.9.2.2 *Local Balance Equations*   Recall that the steady-state solution of the birth and death process is given by:

$$\lambda_0 p_0 = \mu_1 p_1,$$
$$(\lambda_i + \mu_i) = \mu_{i+1} p_{i+1} + \lambda_{i-1} p_{i-1}, i = 1, 2, \ldots,$$
$$\sum_i p_i = 1.$$

For $i = 1$, we obtain $(\lambda_1 + \mu_1)\, p_1 = \mu_2 p_2 + \lambda_0 p_0$. Since we know from the first equation that $\lambda_0\, p_0 = \mu_1 p_1$, this equation becomes:

$$\lambda_1 p_1 = \mu_2 p_2.$$

Similarly, for $i = 2$, we have that $(\lambda_2 + \mu_2)\, p_2 = \mu_3 p_3 + \lambda_1 p_2$. Applying the last result, we obtain:

$$\lambda_2 p_2 = \mu_3 p_3.$$

Repeated application of this method yields the general result

$$\lambda_i p_i = \mu_{i+1} p_{i+1} \quad i = 0, 1, \ldots.$$

This result states that when the process is in the steady state, the rate at which it makes a transition from state i to state $i + 1$, which we refer to as the rate of flow from state i to state $i + 1$, is equal to the rate of flow from state $i + 1$ to state i. This property is referred to as *local balance* condition. Direct application of the property allows us to solve for the steady-state probabilities recursively as follows:

$$p_{i+1} = \frac{\lambda_i}{\mu_{i+1}} p_i$$

$$= \frac{\lambda_i \lambda_{i-1}}{\mu_{i+1} \mu_i} p_{i-1}$$

$$\ldots$$

$$= \frac{\lambda_i \lambda_{i-1} \ldots \lambda_0}{\mu_{i+1} \mu_i \ldots \mu_1} p_0,$$

$$1 = p_0 \left[1 + \sum_{i=1}^{\infty} \frac{\lambda_i \lambda_{i-1} \ldots \lambda_0}{\mu_{i+1} \mu_i \ldots \mu_1} \right],$$

$$p_0 = \left[1 + \sum_{i=1}^{\infty} \frac{\lambda_i \lambda_{i-1} \ldots \lambda_0}{\mu_{i+1} \mu_i \ldots \mu_1} \right]^{-1},$$

$$p_i = \frac{\lambda_i \lambda_{i-1} \ldots \lambda_0}{\mu_{i+1} \mu_i \ldots \mu_1} \left[1 + \sum_{i=1}^{\infty} \frac{\lambda_i \lambda_{i-1} \ldots \lambda_0}{\mu_{i+1} \mu_i \ldots \mu_1} \right]^{-1}, \quad i \geq 1.$$

When $\lambda_i = \lambda$ for all i and $\mu_i = \mu$ for all i, we obtain the result:

$$p_0 = \left[1 + \sum_{i=1}^{\infty} \left(\frac{\lambda}{\mu} \right)^i \right]^{-1}.$$

The sum converges if and only if $\lambda/\mu < 1$, which is equivalent to the condition that $\lambda < \mu$. Under this condition we obtain the solutions:

$$p_0 = 1 - \frac{\lambda}{\mu},$$

$$p_i = \left(1 - \frac{\lambda}{\mu} \right) \left(\frac{\lambda}{\mu} \right)^i, \quad i \geq 1.$$

In Chapter 3, we will refer to this special case of the birth and death process as an M/M/1 queueing system.

2.10 GAUSSIAN PROCESSES

Gaussian processes are important in many ways. First, many physical problems are the results of adding large numbers of independent random variables. According to the central limit theorem, such sums of random variables are essentially normal (or Gaussian) random variables. Also, the analysis of many systems is simplified if they are assumed to be Gaussian processes because of the properties of Gaussian processes. For example, noise in communication systems is usually modeled as a Gaussian process. Similarly, noise voltages in resistors are modeled as Gaussian processes.

A stochastic process $\{X(t), t \in T\}$ is defined to be a Gaussian process if and only if for any choice of n real coefficients $a_1, a_2, \ldots, a_n$ and choice of n time instants $t_1, t_2, \ldots, t_n$ in the index set T, the random variable $a_1 X(t_1) + a_2 X(t_2) + \ldots + a_n X(t_n)$ is a Gaussian (or normal) random variable. That is, $\{X(t), t \in T\}$ is a Gaussian process if any finite linear combination of the $X(t)$ is a normally distributed random variable. This definition implies that the random variables $X(t_1), X(t_2), \ldots, X(t_n)$ have a jointly normal PDF; that is,

$$f_{X(t_1)X(t_2)\ldots X(t_n)}(x_1, x_2, \ldots, x_n) = \frac{1}{(2\pi)^{n/2} |\mathbf{C_{XX}}|^{1/2}} \exp\left[-\frac{(\mathbf{x}-\mathbf{\mu_X})^{\mathrm{T}} \mathbf{C_{XX}^{-1}}(\mathbf{x}-\mathbf{\mu_X})}{2}\right],$$

where $\mu_\mathbf{X}$ is the vector of the mean functions of the $X(t_k)$, $\mathbf{C_{XX}}$ is the matrix of the autocovariance functions, X is the vector of the $X(t_k)$, and T denotes the transpose operation. That is,

$$\mathbf{\mu_X} = \begin{bmatrix} \mu_X(t_1) \\ \mu_X(t_2) \\ \ldots \\ \mu_X(t_n) \end{bmatrix} \quad \mathbf{X} = \begin{bmatrix} X(t_1) \\ X(t_2) \\ \ldots \\ X(t_n) \end{bmatrix},$$

$$\mathbf{C_{XX}} = \begin{bmatrix} C_{XX}(t_1, t_1) & C_{XX}(t_1, t_2) & \ldots & C_{XX}(t_1, t_n) \\ C_{XX}(t_2, t_1) & C_{XX}(t_2, t_2) & \ldots & C_{XX}(t_2, t_n) \\ \ldots & \ldots & \ldots & \ldots \\ C_{XX}(t_n, t_1) & C_{XX}(t_n, t_2) & \ldots & C_{XX}(t_n, t_n) \end{bmatrix}.$$

If the $X(t_k)$ are mutually uncorrelated, then:

$$C_{XX}(t_i, t_j) = \begin{cases} \sigma_X^2 & i = j \\ 0 & \text{otherwise.} \end{cases}$$

The autocovariance matrix and its inverse become:

$$C_{XX} = \begin{bmatrix} \sigma_X^2 & 0 & \cdots & 0 \\ 0 & \sigma_X^2 & \cdots & 0 \\ \cdots & \cdots & \cdots & \cdots \\ 0 & 0 & \cdots & \sigma_X^2 \end{bmatrix} \qquad C_{XX}^{-1} = \begin{bmatrix} \dfrac{1}{\sigma_X^2} & 0 & \cdots & 0 \\ 0 & \dfrac{1}{\sigma_X^2} & \cdots & 0 \\ \cdots & \cdots & \cdots & \cdots \\ 0 & 0 & \cdots & \dfrac{1}{\sigma_X^2} \end{bmatrix}.$$

Thus, we obtain

$$(x - \mu_X)^T C_{XX}^{-1} (x - \mu_X) = \sum_{k=1}^{N} \frac{[x_k - \mu_X(t_k)]^2}{\sigma_X^2},$$

$$f_{X(t_1)X(t_2)\ldots X(t_n)}(x_1, x_2, \ldots, x_n) = \frac{1}{(2\pi\sigma_X^2)^{n/2}} \exp\left[-\frac{1}{2}\sum_{k=1}^{n}\left\{\frac{x_k - \mu_X(t_k)}{\sigma_X}\right\}^2\right].$$

If in addition to being mutually uncorrelated the random variables $X(t_1)$, $X(t_2), \ldots, X(t_n)$ have different variances such that $\text{Var}(X(t_k)) = \sigma_k^2, (1 \leq k \leq n)$, then the covariance matrix and the joint PDF are given by:

$$C_{XX} = \begin{bmatrix} \sigma_1^2 & 0 & \cdots & 0 \\ 0 & \sigma_2^2 & \cdots & 0 \\ \cdots & \cdots & \cdots & \cdots \\ 0 & 0 & \cdots & \sigma_n^2 \end{bmatrix},$$

$$f_{X(t_1)X(t_2)\ldots X(t_n)}(x_1, x_2, \ldots, x_n) = \frac{1}{(2\pi)^{n/2}\left(\prod_{k=1}^{n}\sigma_k\right)} \exp\left[-\frac{1}{2}\sum_{k=1}^{n}\left\{\frac{x_k - \mu_X(t_k)}{\sigma_k}\right\}^2\right],$$

which implies that $X(t_1), X(t_2), \ldots, X(t_n)$ are also mutually independent. We list three important properties of Gaussian processes:

1. A Gaussian process that is a WSS process is also a strict-sense stationary process.
2. If the input to a linear system is a Gaussian process, then the output is also a Gaussian process.
3. If the input $X(t)$ to a linear system is a zero-mean Gaussian process, the output process $Y(t)$ is also a zero-mean process. The proof of this property is as follows:

$$Y(t) = \int_{-\infty}^{\infty} h(u) X(t-u) d\tau,$$

$$E[Y(t)] = E\left[\int_{-\infty}^{\infty} h(u) X(t-u) du\right] = \int_{-\infty}^{\infty} h(u) E[X(t-u)] du = 0.$$

Example 2.9: A WSS Gaussian random process has an autocorrelation function:

$$R_{XX}(\tau) = 6e^{-|\tau|/2}.$$

Determine the covariance matrix of the random variables $X(t)$, $X(t+1)$, $X(t+2)$, and $X(t+3)$.

Solution: First, note that:

$$E[X(t)] = \mu_X(t) = \pm\sqrt{\lim_{|\tau|\to\infty}\{R_{XX}(\tau)\}} = 0.$$

Let $X_1 = X(t)$, $X_2 = X(t+1)$, $X_3 = X(t+2)$, $X_4 = X(t+3)$. Then the elements of the covariance matrix are given by:

$$C_{ij} = Cov(X_i, X_j) = E[(X_i - \mu_{X_i})(X_j - \mu_{X_j})] = E[X_i X_j] = R_{ij} = R_{XX}(i, j)$$
$$= R_{XX}(j-i) = 6e^{-|j-i|/2},$$

where R_{ij} is the i–jth element of the autocorrelation matrix. Thus,

$$C_{XX} = R_{XX} = \begin{bmatrix} 6 & 6e^{-1/2} & 6e^{-1} & 6e^{-3/2} \\ 6e^{-1/2} & 6 & 6e^{-1/2} & 6e^{-1} \\ 6e^{-1} & 6e^{-1/2} & 6 & 6e^{-1/2} \\ 6e^{-3/2} & 6e^{-1} & 6e^{-1/2} & 6 \end{bmatrix}.$$

2.11 PROBLEMS

2.1 Cars arrive from the northbound section of an intersection in a Poisson manner at the rate of λ_N cars per minute and from the eastbound section in a Poisson manner at the rate of λ_E cars per minute.

   a. Given that there is currently no car at the intersection, what is the probability that a northbound car arrives before an eastbound car?

   b. Given that there is currently no car at the intersection, what is the probability that the fourth northbound car arrives before the second eastbound car?

2.2 Suppose $X(t)$ is a Gaussian random process with a mean $E[X(t)] = 0$ and autocorrelation function $R_{XX}(\tau) = e^{-|\tau|}$. Assume that the random variable A is defined as follows:

$$A = \int_0^1 X(t)\,dt.$$

Determine the following:

   a. $E[A]$

   b. σ_A^2

2.3 Suppose $X(t)$ is a Gaussian random process with a mean $E[X(t)] = 0$ and autocorrelation function $R_{XX}(\tau) = e^{-|\tau|}$. Assume that the random variable A is defined as follows:

$$A = \int_0^B X(t)\,dt,$$

where B is a uniformly distributed random variable with values between 1 and 5 and is independent of the random process $X(t)$. Determine the following:

a. $E[A]$

b. σ_A^2

2.4 Consider a machine that is subject to failure and repair. The time to repair the machine when it breaks down is exponentially distributed with mean $1/\mu$. The time the machine runs before breaking down is also exponentially distributed with mean $1/\lambda$. When repaired, the machine is considered to be as good as new. The repair time and the running time are assumed to be independent. If the machine is in good condition at time 0, what is the expected number of failures up to time t?

2.5 The Merrimack Airlines company runs a commuter air service between Manchester, New Hampshire, and Cape Cod, Massachusetts. Because the company is a small one, there is no set schedule for their flights, and no reservation is needed for the flights. However, it has been determined that their planes arrive at the Manchester airport according to a Poisson process with an average rate of two planes per hour. Gail arrived at the Manchester airport and had to wait to catch the next flight.

a. What is the mean time between the instant Gail arrived at the airport until the time the next plane arrived?

b. What is the mean time between the arrival time of the last plane that took off from the Manchester airport before Gail arrived and the arrival time of the plane that she boarded?

2.6 Victor is a student who is conducting experiments with a series of light bulbs. He started with 10 identical light bulbs, each of which has an exponentially distributed lifetime with a mean of 200 h. Victor wants to know how long it will take until the last bulb burns out (or fails). At noontime, he stepped out to get some lunch with six bulbs still on. Assume that he came back and found that none of the six bulbs has failed.

a. After Victor came back, what is the expected time until the next bulb failure?

b. What is the expected length of time between the fourth bulb failure and the fifth bulb failure?

2.7 A machine has three components labeled 1, 2, and 3, whose times between failure are exponentially distributed with mean $1/\lambda_1$, $1/\lambda_2$ and $1/\lambda_3$, respectively. The machine needs all three components to work, thus when a component fails the machine is shut down until the component is repaired and the machine is brought up again. When repaired, a component is considered to be as good as new. The time to repair component 1 when it fails is exponentially distributed with mean $1/\mu_1$. The time to repair component 2 when it fails is constant at $1/\mu_2$, and the time to repair component 3 when it fails is a third-order Erlang random variable with parameter μ_3.

a. What fraction of time is the machine working?

b. What fraction of time is component 2 being repaired?

c. What fraction of time is component 3 idle but has not failed?

d. Given that Bob arrived when component 1 was being repaired, what is the expected time until the machine is operational again?

2.8 Customers arrive at a taxi depot according to a Poisson process with rate λ. The dispatcher sends for a taxi where there are N customers waiting at the station. It takes M units of time for a taxi to arrive at the depot. When it arrives, the taxi picks up all waiting customers. The taxi company incurs a cost at a rate of nk per unit time whenever n customers are waiting. What is the steady-state average cost that the company incurs?

2.9 Consider a Markov chain with the following transition probability matrix:

$$P = \begin{bmatrix} 0.6 & 0.2 & 0.2 \\ 0.3 & 0.4 & 0.3 \\ 0.0 & 0.3 & 0.7 \end{bmatrix}.$$

a. Give the state transition diagram.

b. Obtain the limiting state probabilities.

c. Given that the process is currently in state 1, what is the probability that it will be in state 2 at the end of the third transition?

3

ELEMENTARY QUEUEING THEORY

3.1 INTRODUCTION

A queue is a waiting line. Queues arise in many of our daily activities. For example, we join a queue to buy stamps at the post office, to cash checks or deposit money at the bank, to pay for groceries at the grocery store, to purchase tickets for movies or games, or to get a table at the restaurant. This chapter discusses a class of queueing systems called Markovian queueing systems. They are characterized by the fact that either the service times are exponentially distributed or customers arrive at the system according to a Poisson process, or both. The emphasis in this chapter is on the steady-state analysis with limited discussion on transient analysis.

3.2 DESCRIPTION OF A QUEUEING SYSTEM

In a queueing system, *customers* from a specified *population* arrive at a *service facility* to receive service. The service facility has one or more *servers* who attend to arriving customers. If a customer arrives at the facility when all the servers are busy attending to earlier customers, the arriving customer joins the queue until a server is free. After a customer has been served, he leaves the system and will not join the queue again. That is, service with feedback is not allowed. We consider systems that obey the *work conservation rule*: A

Fundamentals of Stochastic Networks, First Edition. Oliver C. Ibe.
© 2011 John Wiley & Sons, Inc. Published 2011 by John Wiley & Sons, Inc.

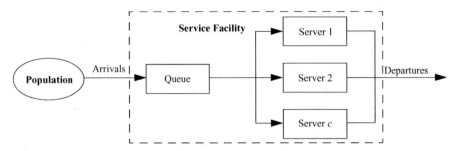

Figure 3.1 Components of a queueing system.

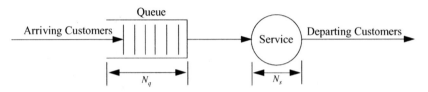

Figure 3.2 The queueing process.

server cannot be idle when there are customers to be served. Figure 3.1 illustrates the different components of a queueing system.

When a customer arrives at a service facility, a server commences service
on the customer if the server is currently idle. Otherwise, the customer joins
a queue that is attended to in accordance with a specified service policy such
as first-come, first-served (FCFS), last-come, first served (LCFS), priority, and
so on. Thus, the time a customer spends waiting for service to begin is dependent on the service policy. Also, since there can be one or more servers in the
facility, more than one customer can be receiving service at the same time. The
following notation is used to represent the random variables associated with
a queueing system:

a. N_q is the number of customers in queue, waiting to be served
b. N_s is the number of customers currently receiving service
c. N is the total number of customers in the system: $N = N_q + N_s$
d. W is the time a customer spends in queue before going to service; W is
   the waiting time
e. X is the time a customer spends in actual service
f. T is the total time a customer spends in the system (also called the
   sojourn time): $T = W + X$

Figure 3.2 is a summary of the queueing process at the service facility.
A queueing system is characterized as follows:

a. *Population*, which is the source of the customers arriving at the service facility. The population can be finite or infinite

b. *Arriving pattern*, which defines the customer interarrival process

c. *Service time distribution*, which defines the time taken to serve each customer

d. *Capacity of the queueing facility*, which can be finite or infinite. If the capacity is finite, customers that arrive when the system is full are lost (or blocked). Thus a finite-capacity system is a blocking system (or a loss system).

e. *Number of servers*, which can be one or more than one. A queueing system with one server is called a *single-server system*; otherwise, it is called a *multiserver system*. A single-server system can serve only one customer at a time while multiserver systems can serve multiple customers simultaneously. In a multiserver system, the servers can be identical, which means that their service rates are identical and it does not matter which server a particular customer receives service from. On the other hand, the servers can be heterogeneous in the sense that some of them provide faster service than others. In this case, the time a customer spends in service depends on which server provides the service. A special case of a multiserver system is the infinite-server system where each arriving customer is served immediately; that is, there is no waiting in queue.

f. *Queueing discipline*, which is also called the *service discipline*. It defines the rule that governs how the next customer to receive service is selected after a customer who is currently receiving service leaves the system. Specific disciplines that can be used include the following:

- FCFS, which means that customers are served in the order they arrived. The discipline is also called first in, first out (FIFO).

- LCFS, which means that the last customer to arrive receives service before those that arrived earlier. The discipline is also called last in, first out (LIFO)

- Service in random order (SIRO), which means that the next customer to receive service after the current customer has finished receiving service will be selected in a probabilistic manner, such as tossing a coin, rolling a die, and so on.

- Priority, which means that customers are divided into ordered classes such that a customer in a higher class will receive service before a customer in a lower class, even if the higher-class customer arrives later than the lower-class customer. There are two types of priority: *preemptive* and *nonpreemptive*. In preemptive priority, the service of a customer currently receiving service is suspended upon the arrival of a higher priority customer; the latter goes straight to receive service. The preempted customer goes in to receive service upon the completion

of service of the higher priority customer, if no higher priority customer arrived while the high-priority customer was being served. How
the service of a preempted customer is continued when the customer
goes to complete his service depends on whether we have a *preemptive
repeat* or *preemptive resume* policy. In preemptive repeat, the customer's service is started from the beginning when the customer enters
to receive service again, regardless of how many times the customer
is preempted. In preemptive resume, the customer's service continues
from where it stopped before being preempted. Under nonpreemptive
priority, an arriving high-priority customer goes to the head of the
queue and waits for the current customer's service to be completed
before he enters to receive service ahead of other waiting lower priority customers.

Thus, the time a customer spends in the system is a function of the above
parameters and service policies.

3.3   THE KENDALL NOTATION

The Kendall notation is a shorthand notation that is used to describe queueing
systems. It is written in the form:

$$A/B/c/D/E/F.$$

- "A" describes the arrival process (or the interarrival time distribution),
  which can be an exponential or nonexponential (i.e., general)
  distribution.
- "B" describes the service time distribution.
- "c" describes the number of servers.
- "D" describes the system capacity, which is the maximum number of
  customers allowed in the system, including those currently receiving
  service; the default value is infinity.
- "E" describes the size of the population from where arrivals are drawn;
  the default value is infinity.
- "F" describes the queueing (or service) discipline. The default is FCFS.
- When default values of D, E, and F are used, we use the notation A/B/c,
  which means a queueing system with infinite capacity, customers arrive
  from an infinite population, and service is in an FCFS manner. Symbols
  traditionally used for A and B are:
  - GI, which stands for general independent interarrival time; it is some-
    times written as G.
  - G, which stands for general service time distribution.

- M, which stands for memoryless (or exponential) interarrival time or service time distribution. Note that an exponentially distributed interarrival time means that customers arrive according to a Poisson process.
- D, which stands for deterministic (or constant) interarrival time or service time distribution.

For example, we can have queueing systems of the following form:

- M/M/1 queue, which is a queueing system with exponentially distributed interarrival time (i.e., Poisson arrivals), exponentially distributed service time, a single server, infinite capacity, customers are drawn from an infinite population, and service is on an FCFS basis.
- M/D/1 queue, which is a queueing system with exponentially distributed interarrival time (i.e., Poisson arrivals), constant service time, a single server, infinite capacity, customers are drawn from an infinite population, and service is on an FCFS basis.
- G/M/3/20 queue, which is a queueing system with a general interarrival time, exponentially distributed service time, three servers, a finite capacity of 20 (i.e., a maximum of 20 customers can be in the system, including the three that can be in service at the same time), customers are drawn from an infinite population, and service is on an FCFS basis.

3.4 THE LITTLE'S FORMULA

The Little's formula (Little 1961) is a statement on the relationship between the mean number of customers in the system, the mean time spent in the system, and the average rate at which customers arrive at the system. Let λ denote the mean arrival rate, $E[N]$ the mean number of customers in the system, $E[T]$ the mean total time spent in the system, $E[N_q]$ the mean number of customers in the queue, and $E[W]$ the mean waiting time. Then, Little's formula states that:

$$E[N] = \lambda E[T],$$

which says that the mean number of customers in the system (including those currently being served) is the product of the average arrival rate and the mean time a customer spends in the system. The formula can also be stated in terms of the number of customers in queue, as follows:

$$E[N_q] = \lambda E[W],$$

which says that the mean number of customers in queue (or waiting to be served) is equal to the product of the average arrival rate and the mean waiting time.

3.5 THE M/M/1 QUEUEING SYSTEM

This is the simplest queueing system in which customers arrive according to a Poisson process to a single-server service facility, and the time to serve each customer is exponentially distributed. The model also assumes the various default values: infinite capacity at the facility, customers are drawn from an infinite population, and service is on an FCFS basis.

Since we are dealing with a system that can increase or decrease by at most one customer at a time, it is a birth and death process with homogeneous birth rate λ and homogeneous death rate μ. This means that the service time is $1/\mu$. Thus, the state transition rate diagram is shown in Figure 3.3.

Let p_n be the limiting state probability that the process is in state n, $n = 0$, $1, 2, \ldots$. Then applying the balance equations we obtain:

$$\lambda p_0 = \mu p_1 \Rightarrow p_1 = \left(\frac{\lambda}{\mu}\right) p_0 = \rho p_0$$

$$(\lambda + \mu) p_1 = \lambda p_0 + \mu p_2 \Rightarrow p_2 = \rho p_1 = \rho^2 p_0$$

$$(\lambda + \mu) p_2 = \lambda p_1 + \mu p_3 \Rightarrow p_3 = \rho p_2 = \rho^3 p_0,$$

where $\rho = \lambda/\mu$. Similarly, it can be shown that:

$$p_n = \rho^n p_0 \quad n = 0, 1, 2, \ldots.$$

Since:

$$\sum_{n=0}^{\infty} p_n = 1 = p_0 \sum_{n=0}^{\infty} \rho^n = \frac{p_0}{1-\rho},$$

we obtain

$$p_0 = 1 - \rho$$

$$p_n = (1-\rho)\rho^n \quad n = 0, 1, 2, \ldots; \rho < 1.$$

Since $\rho = 1 - p_0$, which is the probability that the system is not empty and hence the server is not idle, we call ρ the *server utilization* (or *utilization factor*).

The expected number of customers in the system is given by:

$$E[N] = \sum_{n=0}^{\infty} n p_n = \sum_{n=0}^{\infty} n(1-\rho)\rho^n = (1-\rho)\sum_{n=0}^{\infty} n\rho^n.$$

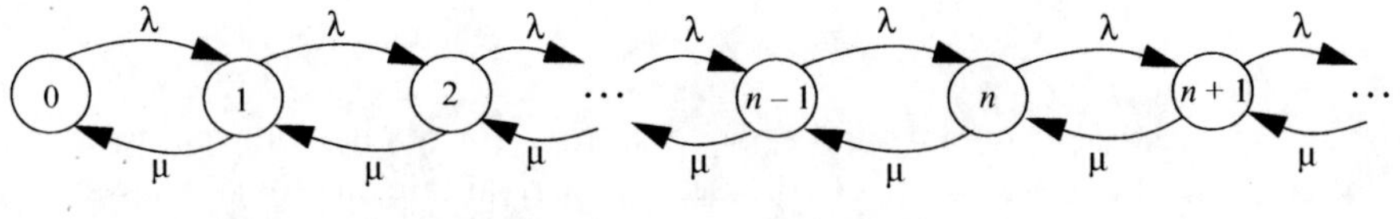

Figure 3.3 State transition rate diagram for M/M/1 queue.

But:

$$\frac{d}{d\rho}\sum_{n=0}^{\infty}\rho^n = \sum_{n=0}^{\infty}\frac{d}{d\rho}\rho^n = \sum_{n=0}^{\infty}n\rho^{n-1} = \frac{1}{\rho}\sum_{n=0}^{\infty}n\rho^n.$$

Thus,

$$\sum_{n=0}^{\infty}n\rho^n = \rho\frac{d}{d\rho}\sum_{n=0}^{\infty}\rho^n = \rho\frac{d}{d\rho}\left(\frac{1}{1-\rho}\right) = \frac{\rho}{(1-\rho)^2}.$$

Therefore,

$$E[N] = (1-\rho)\sum_{n=0}^{\infty}n\rho^n = (1-\rho)\frac{\rho}{(1-\rho)^2} = \frac{\rho}{1-\rho}.$$

We can obtain the mean time in the system from Little's formula as follows:

$$E[T] = E[N]/\lambda = \frac{\lambda/\mu}{\lambda(1-\rho)}$$

$$= \frac{1}{\mu(1-\rho)} = \frac{E[X]}{1-\rho},$$

where the last result follows from the fact that the mean service time is $E[X] = 1/\mu$. Similarly, the mean waiting time and mean number of customers in queue are given by:

$$E[W] = E[T] - E[X] = \frac{E[X]}{1-\rho} - E[X]$$

$$= \frac{\rho E[X]}{1-\rho} = \frac{\rho}{\mu(1-\rho)}$$

$$E[N_q] = \lambda E[W] = \frac{\lambda\rho}{\mu(1-\rho)}$$

$$= \frac{\rho^2}{1-\rho}.$$

Recall that the mean number of customers in service is $E[N_s]$. Using the above results we obtain:

$$E[N_s] = E[N] - E[N_q] = \frac{\rho}{1-\rho} - \frac{\rho^2}{1-\rho} = \rho.$$

Thus, the mean number of customers in service is ρ, the probability that the server is busy.

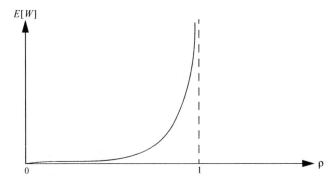

Figure 3.4   Mean waiting time versus server utilization.

Note that the mean waiting time, the mean time in the system, the mean number of customers in the system, and the mean number of customers in queue become extremely large as the server utilization ρ approaches 1. Figure 3.4 illustrates this for the case of the expected waiting time.

Example 3.1: Students arrive at the campus post office according to a Poisson process with an average rate of one student every 4 min. The time required to serve each student is exponentially distributed with a mean of 3 min. There is only one postal worker at the counter and any arriving student who finds the worker busy joins a queue that is served in an FCFS manner.

    a. What is the probability that an arriving student has to wait?
    b. What is the mean waiting time of an arbitrary student?
    c. What is the mean number of waiting students at the post office?

Solution: The above problem is an M/M/1 queue with the following parameters:

$$\lambda = 1/4,$$
$$\mu = 1/3,$$
$$\rho = \lambda/\mu = 3/4 = 0.75.$$

    a. P[arriving student waits] = P[server is busy] = ρ = 0.75
    b. $E[W] = \rho E[X]/(1 - \rho) = (0.75)(3)/(1 - 0.75) = 9$ min
    c. $E[N_q] = \lambda E[W] = (1/4)(9) = 2.25$ students.

Example 3.2: Customers arrive at a checkout counter in a grocery store according to a Poisson process with an average rate of 10 customers per hour. There is only one clerk at the counter and the time to serve each customer is exponentially distributed with a mean of 4 min.

a. What is the probability that a queue forms at the counter?
b. What is the average time a customer spends at the counter?
c. What is the average queue length at the counter?

Solution: This is an M/M/1 queueing system. We must first convert the arrival and service rates to the same unit of customers per minute since service time is in minutes. Thus, the parameters of the system are as follows:

$$\lambda = 10/60 = 1/6,$$
$$\mu = 1/4 = 1/E[X],$$
$$\rho = \lambda/\mu = 2/3.$$

a. P[queue forms] = P[server is busy] = ρ = 2/3
b. Average time at the counter = $E[T] = E[X]/(1 - \rho) = 4/(1/3) = 12\,\text{min}$
c. Average queue length = $E[N_q] = \rho^2/(1 - \rho) = (4/9)/(1/3) = 4/3$ customers

3.5.1 Stochastic Balance

A shortcut method of obtaining the steady-state equations in an M/M/1 queueing system is by means of a flow balance procedure called *local balance*. The idea behind stochastic balance is that in any steady-state condition, the rate at which the process moves from left to right across a "probabilistic wall" is equal to the rate at which it moves from right to left across that wall. For example, Figure 3.5 shows three states $n - 1, n$, and $n + 1$ in the state transition rate diagram of an M/M/1 queue.

The rate at which the process crosses wall A from left to right is λp_{n-1}, which is true because this can only happen when the process is in state $n - 1$. Similarly, the rate at which the process crosses wall A from right to left is μp_n. By stochastic balance we mean that $\lambda p_{n-1} = \mu p_n$ or $p_n = (\lambda/\mu)p_{n-1} = \rho p_{n-1}$, which is the result we obtained earlier in the analysis of the system.

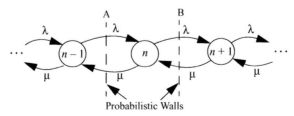

Figure 3.5   Local balance concept.

3.5.2 Total Time and Waiting Time Distributions of the M/M/1 Queueing System

Consider a tagged customer, say customer k, who arrives at the system and finds it in state n. If $n > 0$, then because of the fact that the service time X is exponentially distributed, the service time of the customer receiving service when the tagged customer arrives "starts from scratch." Thus, the total time that the tagged customer spends in the system is given by:

$$T_k = \begin{cases} X_k + X_1 + X_2 + \ldots + X_n & n > 0 \\ X_k & n = 0. \end{cases}$$

Since the X_k are identically distributed, the s-transform of the probability density function (PDF) of T_k is given by:

$$M_{T|n}(s|n) = \left\{ M_X(s)^{n+1} \right\} \quad n = 0, 1, \ldots.$$

Thus, the unconditional total time in the system is given by:

$$M_T(s) = \sum_{n=0}^{\infty} M_{T|n}(s|n) p_n = \sum_{n=0}^{\infty} \{M_X(s)\}^{n+1} p_n = M_X(s) \sum_{n=0}^{\infty} \{M_X(s)\}^n (1-\rho)\rho^n$$

$$= (1-\rho) M_X(s) \sum_{n=0}^{\infty} \{\rho M_X(s)\}^n = \frac{(1-\rho) M_X(s)}{1 - \rho M_X(s)}.$$

Since $M_X(s) = \mu/(s + \mu)$, we obtain the following result:

$$M_T(s) = \frac{\mu(1-\rho)}{s + \mu(1-\rho)}.$$

This shows that T is an exponentially distributed random variable with mean $1/\mu(1 - \rho)$, which agrees with the value of $E[T]$ that we obtained earlier. Similarly, the waiting time distribution can be obtained by considering the experience of the tagged customer in the system, as follows:

$$W_k = \begin{cases} X_1 + X_2 + \ldots + X_n & n > 0 \\ 0 & n = 0. \end{cases}$$

Thus, the s-transform of the PDF of W_k, given that there are n customers when the tagged customer arrives, is given by:

$$M_{W|n}(s|n) = \begin{cases} \{M_X(s)\}^n & n = 1, 2, \ldots \\ 1 & n = 0. \end{cases}$$

Therefore,

$$M_W(s) = p_0 + \sum_{n=1}^{\infty} M_{W|n}(s|n) p_n = p_0 + \sum_{n=1}^{\infty} \{M_X(s)\}^n p_n$$

$$= (1-\rho) + \sum_{n=1}^{\infty} \{M_X(s)\}^n (1-\rho)\rho^n$$

$$= (1-\rho) + (1-\rho) \sum_{n=1}^{\infty} \{\rho M_X(s)\}^n = (1-\rho)\left[1 + \left\{\frac{1}{1-\rho M_X(s)} - 1\right\}\right]$$

$$= (1-\rho)\left\{1 + \frac{\rho M_X(s)}{1-\rho M_X(s)}\right\} = (1-\rho)\left\{\frac{1}{1-\rho M_X(s)}\right\} = (1-\rho)\left\{1 + \frac{\rho\mu}{s+\mu(1-\rho)}\right\}$$

$$= (1-\rho) + \frac{\rho\mu(1-\rho)}{s+\mu(1-\rho)}.$$

Thus, the PDF and cumulative distribution function (CDF) of W are given by:

$$f_W(w) = (1-\rho)\delta(w) + \rho\mu(1-\rho)e^{-\mu(1-\rho)w} = (1-\rho)\delta(w) + \lambda(1-\rho)e^{-\mu(1-\rho)w}$$
$$F_W(w) = P[W \le w] = (1-\rho) + \rho\{1 - e^{-\mu(1-\rho)w}\} = 1 - \rho e^{-\mu(1-\rho)w},$$

where $\delta(w)$ is the impulse function. Observe that the mean waiting time is given by:

$$E[W] = \int_0^{\infty} \{1 - F_W(w)\} dw = \int_0^{\infty} \rho e^{-\mu(1-\rho)w} dw = \frac{\rho}{\mu(1-\rho)},$$

which was the result we obtained earlier.

3.6 EXAMPLES OF OTHER M/M QUEUEING SYSTEMS

The goal of this section is to describe some relatives of the M/M/1 queueing systems without rigorously analyzing them as we did for the M/M/1 queueing system. These systems can be used to model different human behaviors and they include:

a. *Blocking* from entering a queue, which is caused by the fact that the system has a finite capacity and a customer arrives when the system is full.

b. *Defections* from a queue, which can be caused by the fact that a customer has spent too much time in queue and leaves without receiving service out of frustration. Defection is also called *reneging* in queueing theory.

c. *Jockeying* for position among many queues, which can arise when in a multiserver system, each server has its own queue and some customer in one queue notices that another queue is being served faster than his own, thus he leaves his queue and moves into the supposedly faster queue.

d. *Balking* before entering a queue, which can arise if the customer perceives the queue to be too long and chooses not to join it at all

e. *Bribing* for queue position, which is a form of dynamic priority because a customer pays some "bribe" to improve his position in the queue. Usually the more bribes he pays, the better position he gets.

f. *Cheating* for queue position, which is different from bribing because in cheating, the customer uses trickery rather than his personal resources to improve his position in the queue

g. *Bulk service*, which can be used to model table assignments in restaurants. For example, a queue at a restaurant may appear to be too long, but in actual fact when a table is available (i.e., a server is ready for the next customer), it can be assigned to a family of four, which is identical to serving the four people in queue together.

h. *Batch arrival*, which can be used to model how friends arrive in groups at a movie theater, concert show or a ball game, or how families arrive at a restaurant. Thus, the number of customers in each arriving batch can be modeled by some probabilistic law.

From this list, it can be seen that queueing theory is a very powerful modeling tool that can be applied to all human activities and hence all walks of life. In the following sections we describe some of the different queueing models that are based on Poisson arrivals and exponentially distributed service times.

3.6.1   The M/M/c Queue: The c-Server System

In this scheme there are *c* identical servers. When a customer arrives he is randomly assigned to one of the idle servers until all servers are busy when a single queue is formed. Note that if a queue is allowed to form in front of each server, then we have an M/M/1 queue with modified arrival since customers join a server's queue in some probabilistic manner. In the single queue case, we assume that there is an infinite capacity and service is based on FCFS policy.

The service rate in the system is dependent on the number of busy servers. If only one server is busy, the service rate is μ; if two servers are busy, the service rate is 2μ; and so on until all servers are busy when the service rate becomes $c\mu$. Thus, until all servers are busy the system behaves like a heterogeneous queueing system in which the service rate in each state is different. When all servers are busy, it behaves like a homogeneous queueing system in which the service rate is the same in each state. The state transition rate diagram of the system is shown in Figure 3.6.

The service rate is:

$$\mu_n = \begin{cases} n\mu & 1 \le n < c \\ c\mu & n \ge c. \end{cases}$$

If we define $\rho = \lambda/c\mu$, then using stochastic balance equations it can be shown that the limiting state probability of being in state n is given by:

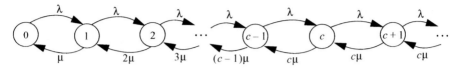

Figure 3.6 State transition rate diagram for M/M/c queue.

$$p_n = \left(\frac{\lambda}{\mu}\right)^n \left(\frac{1}{n!}\right) p_0 = \frac{(c\rho)^n}{n!} p_0 \quad n = 0, 1, \dots, c,$$

$$p_n = \frac{(\lambda/\mu)^n}{c! c^{n-c}} p_0 = \frac{c^c \rho^n}{c!} p_0 \quad n \geq c,$$

$$\sum_{n=0}^{\infty} p_n = 1 \Rightarrow p_0 = \left[\sum_{n=0}^{c-1} \frac{(c\rho)^n}{n!} + \frac{(c\rho)^c}{c!}\left(\frac{1}{1-\rho}\right)\right]^{-1} \quad \rho < 1.$$

Note that queues can only form when the process is in state c or any state higher than c. Thus, arriving customers who see the system in any state less than c do not have to wait. The probability that a queue is formed is given by:

$$P_Q = \sum_{n=c}^{\infty} p_n = p_0 \sum_{n=c}^{\infty} \frac{c^c \rho^n}{c!} = p_0 \frac{(c\rho)^c}{c!} \sum_{n=c}^{\infty} \rho^{n-c} = \frac{(c\rho)^c}{c!}\left(\frac{1}{1-\rho}\right) p_0.$$

The expected number of customers in queue is given by:

$$E[N_q] = \sum_{n=c}^{\infty} (n-c) p_n = p_0 \frac{(c\rho)^c}{c!} \sum_{n=c}^{\infty} (n-c)\rho^{n-c} = \frac{(c\rho)^c}{c!} \frac{\rho}{(1-\rho)^2} p_0$$

$$= \frac{\rho}{1-\rho} P_Q.$$

From Little's formula, the mean waiting time is given by:

$$E[W] = \frac{E[N_q]}{\lambda} = \frac{\rho}{\lambda(1-\rho)} P_Q.$$

Finally, the expected time in the system and the expected number of customers in the system are given, respectively, by:

$$E[T] = E[W] + \frac{1}{\mu} = \frac{\rho}{\lambda(1-\rho)} P_Q + \frac{1}{\mu},$$

$$E[N] = \lambda E[T] = \frac{\rho}{1-\rho} P_Q + c\rho.$$

Example 3.3: Students arrive at a checkout counter in the college cafeteria according to a Poisson process with an average rate of 15 students per hour.

There are two cashiers at the counter and they provide identical service to students. The time to serve a student by either cashier is exponentially distributed with a mean of 3 min. Students who find both cashiers busy on their arrival join a single queue. What is the probability that an arriving student does not have to wait?

Solution: This is an M/M/2 queueing problem with the following parameters with λ and μ in students per minute:

$$\lambda = 15/60 = 1/4,$$

$$\mu = 1/3,$$

$$p_0 = \cfrac{1}{1 + \cfrac{\lambda}{\mu} + \left(\cfrac{\lambda}{\mu}\right)^2 \left[\cfrac{2\mu}{2(2\mu - \lambda)}\right]} = \frac{5}{11},$$

$$p_1 = \left(\frac{\lambda}{\mu}\right) p_0 = \left(\frac{3}{4}\right)\left(\frac{5}{11}\right) = \frac{15}{44}.$$

An arriving student does not have to wait if he finds the system either empty or with only one server busy. The probability of this event is $p_0 + p_1 = 35/44$.

3.6.2   The M/M/1/K Queue: The Single-Server Finite-Capacity System

In this system, arriving customers who see the system in state K are lost. The state transition rate diagram is shown in Figure 3.7.

Using the stochastic balance equations, it can be shown that the steady-state probability that the process is in state n is given by:

$$p_n = \left(\frac{\lambda}{\mu}\right)^n p_0 \quad n = 0, 1, 2, \ldots, K,$$

$$\sum_{n=0}^{K} p_n = 1 \Rightarrow p_0 = \frac{1 - (\lambda/\mu)}{1 - (\lambda/\mu)^{K+1}},$$

where $\lambda < \mu$. It can also be shown that the mean number of customers in the system is given by:

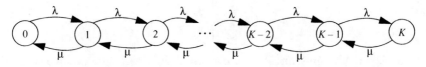

Figure 3.7   State transition rate diagram for M/M/1/K queue.

$$E[N] = \frac{\lambda/\mu}{1-(\lambda/\mu)} - \frac{(K+1)(\lambda/\mu)^{K+1}}{1-(\lambda/\mu)^{K+1}},$$

$$E[N_q] = E[N] - (1 - p_0).$$

Note that all the traffic reaching the system does not enter the system because customers are not allowed into the system when there are already K customers in the system. That is, if we define $\rho = \lambda/\mu$, then customers are turned away with probability:

$$p_K = \frac{1-(\lambda/\mu)}{1-(\lambda/\mu)^{K+1}} \left(\frac{\lambda}{\mu}\right)^K = \frac{\rho^K(1-\rho)}{1-\rho^{K+1}}.$$

Thus, we define the *actual rate* at which customers arrive into the system, λ_a, as:

$$\lambda_a = \lambda(1 - p_K).$$

We can then apply Little's formula to obtain

$$E[T] = E[N]/\lambda_a,$$

$$E[W] = E[N_q]/\lambda_a.$$

Example 3.4: Each morning, people arrive at Ed's garage to fix their cars. Ed's garage can only accommodate four cars. Anyone arriving when there are already four cars in the garage has to go away without leaving his car for Ed to fix. If Ed's customers arrive according to a Poisson process with a rate of one customer per hour and the time it takes Ed to service a car is exponentially distributed with a mean of 45 min, what is the probability that an arriving customer finds Ed idle? What is the probability that an arriving customer leaves without getting his car fixed?

Solution: This is an M/M/1/4 queue with the following parameters:

$$\lambda = 1,$$
$$\mu = 60/45 = 4/3,$$
$$\lambda/\mu = 3/4,$$
$$p_0 = \frac{1-(\lambda/\mu)}{1-(\lambda/\mu)^5} = \frac{1-(3/4)}{1-(3/4)^5} = \frac{0.25}{0.7627} = 0.3278.$$

a. The probability that an arriving customer finds Ed idle is $p_0 = 0.3278$.
b. The probability that a customer leaves without fixing his car is the probability that he finds the garage full when he arrived, which is $p_4 = (\lambda/\mu)^4 p_0 = 0.1037$

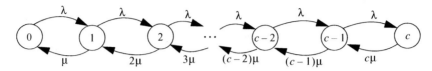

Figure 3.8   State transition rate diagram for M/M/c/c queue.

3.6.3   The M/M/c/c Queue: The c-Server Loss System

This is a very useful model in telephony. It is used to model calls arriving at a telephone switchboard, which usually has a finite capacity. It is assumed that the switchboard can support a maximum of c simultaneous calls (i.e., it has a total of c channels available). Any call that arrives when all c channels are busy will be lost. This is usually referred to as the *blocked-calls-lost* model. The state transition rate diagram is shown in Figure 3.8.

Using the stochastic balance equations, it can be shown that the steady-state probability that the process is in state n is given by:

$$p_n = \frac{(\lambda/\mu)^n /n!}{\displaystyle\sum_{k=0}^{c} (\lambda/\mu)^k /k!} \qquad 0 \le n \le c.$$

The probability that the process is in state c, p_c, is called the *Erlang's loss formula*, which is given by:

$$p_c = \frac{(\lambda/\mu)^c /c!}{\displaystyle\sum_{k=0}^{c} (\lambda/\mu)^k /k!} = \frac{(c\rho)^c /c!}{\displaystyle\sum_{k=0}^{c} (c\rho)^k /k!},$$

where $\rho = \lambda/c\mu$ is the utilization factor of the system.

As in the M/M/1/K queueing system, not all traffic enters the system. The actual average arrival rate into the system is:

$$\lambda_a = \lambda(1 - p_c).$$

Since no customer is allowed to wait, $E[W]$ and $E[N_q]$ are both zero. However, the mean number of customers in the system is:

$$E[N] = \sum_{n=0}^{c} np_n = p_0 \sum_{n=1}^{c} n\frac{(\lambda/\mu)^n}{n!} = (\lambda/\mu)p_0 \sum_{n=1}^{c} \frac{(\lambda/\mu)^{n-1}}{(n-1)!} = (\lambda/\mu)p_0 \sum_{n=0}^{c-1} \frac{(\lambda/\mu)^n}{n!}$$

$$= (\lambda/\mu)[1 - p_c].$$

By Little's formula,

$$E[T] = E[N]/\lambda_a = 1/\mu.$$

This confirms that the mean time a customer admitted into the system spends in the system is the mean service time.

Example 3.5: Bob established a dial-up service for Internet access. As a small businessman, Bob can only support four channels for his customers. Any of Bob's customers who want access to the Internet when all four channels are busy are blocked and will have to hang up. Bob's studies indicate that customers arrive for Internet access according to a Poisson process with an average rate of eight calls per hour and the duration of each call is exponentially distributed with a mean of 10 min. If Jay is one of Bob's customers, what is the probability that on this particular dial-up attempt he cannot gain access to the Internet?

Solution: This is an example of M/M/4/4 queueing system. The parameters of the model are as follows:

$$\lambda = 8/60 = 2/15,$$
$$\mu = 1/10,$$
$$\rho = \lambda/4\mu = 1/3,$$
$$c\rho = 4/3.$$

The probability that Jay is blocked is the probability that he dials up when the process is in state 4. This is given by:

$$p_4 = \frac{(c\rho)^4/4!}{\sum_{k=0}^{4}(c\rho)^k/k!} = \frac{(4/3)^4/24}{1+(4/3)+\frac{(4/3)^2}{2}+\frac{(4/3)^3}{6}+\frac{(4/3)^4}{24}}$$

$$= \frac{0.1317}{1+1.3333+0.8889+0.3951+0.1317}$$
$$= 0.0351.$$

3.6.4 The M/M/1//K Queue: The Single Server Finite Customer Population System

In the previous examples we assumed that the customers are drawn from an infinite population because the arrival process has a Poisson distribution. Assume that there are K potential customers in the population. An example is where we have a total of K machines that can be either operational or down, needing a serviceman to fix them. If we assume that the customers act independently of each other and that given that a customer has not yet come to the service facility, the time until he comes to the facility is exponentially distributed with mean $1/\lambda$, then the number of arrivals when n customers are already in the service facility is governed by a heterogeneous Poisson process with mean $\lambda(K - n)$. When $n = K$, there is no more customer left to draw from,

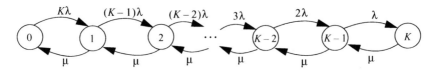

Figure 3.9 State transition rate diagram for M/M/1//K queue.

which means that the arrival rate becomes zero. Thus, the state transition rate diagram is as shown in Figure 3.9.

The arrival rate when the process is in state n is:

$$\lambda_n = \begin{cases} (K-n)\lambda & 0 \le n < K \\ 0 & n > K. \end{cases}$$

Using the stochastic balance equations, it can be shown that the steady-state probabilities are given by:

$$p_n = n!\left(\frac{\lambda}{\mu}\right)^n p_0 = n!\rho^n p_0 \quad n = 0, 1, 2, \dots, K,$$

$$\sum_{n=0}^{K} p_n = 1 \Rightarrow p_0 = \left[\sum_{n=0}^{K} n!\rho^n\right]^{-1},$$

where $\rho = \lambda/\mu$. Other finite-population schemes can easily be derived from the above models. For example, we can obtain the state transition rate diagram for the c-server finite population system with population $K > c$ by combining the arriving process on the M/M/1//K queueing system with the service process of the M/M/c queueing system.

Example 3.6: A small organization has three old PCs, each of which can be working (or operational) or down. When any PC is working, the time until it fails is exponentially distributed with a mean of 10 h. When a PC fails, the repairman immediately commences servicing it to bring it back to the operational state. The time to service each failed PC is exponentially distributed with a mean of 2 h. If there is only one repairman in the facility and the PCs fail independently, what is the probability that the organization has only two PCs working?

Solution: This is an M/M/1//3 queueing problem in which the arrivals are PCs that have failed and the single server is the repairman. Thus, when the process is in state 0 all PCs are working; when it is in state 1, two PCs are working; when it is in state 2, only one PC is working; and when it is in state 3, all PCs are down, awaiting repair. The parameters of the problem are:

$$\lambda = 1/10,$$
$$\mu = 1/2,$$
$$\rho = \lambda/\mu = 0.2,$$

$$p_0 = \left[\sum_{n=0}^{3} n! \rho^n\right]^{-1} = \frac{1}{1 + 0.2 + 2(0.2)^2 + 6(0.2)^3} = 0.7530.$$

As stated earlier, the probability that two computers are working is the probability that the process is in state 1, which is given by:

$$p_1 = \rho p_0 = (0.2)(0.7530) = 0.1506.$$

3.7 M/G/1 QUEUE

In this system, customers arrive according to a Poisson process with rate λ and are served by a single server with a general service time X whose PDF is $f_X(x)$, $x \geq 0$, mean is $E[X]$, second moment is $E[X^2]$, and variance is σ_X^2. The capacity of the system is infinite and customers are served on an FCFS basis. Thus, the service time distribution does not have the memoryless property of the exponential distribution, and the number of customers in the system time t, $N(t)$, is not a Poisson process. A more appropriate description of the state at time t includes both $N(t)$ and the residual life of the service time of the current customer. That is, if R denotes the residual life of the current service, then the set of pairs $\{(N, R)\}$ provides the description of the state space. Thus, we have a two-dimensional state space, which is a somewhat complex way to proceed with the analysis. However, the analysis is simplified if we can identify those points in time where the state is easier to describe. Such points are usually chosen to be those time instants at which customers leave the system, which means that $R = 0$.

To obtain the steady-state analysis of the system, we proceed as follows. Consider the instant the kth customer arrives at the system. Assume that the ith customer was receiving service when the kth customer arrived. Let R_i denote the residual life of the service time of the ith customer at the instant the kth customer arrived, as shown in Figure 3.10.

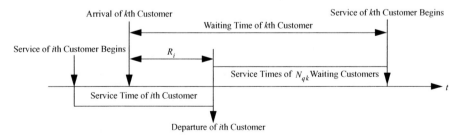

Figure 3.10 Service experience of the ith customer in M/G/1 queue.

Assume that N_{qk} customers were waiting when the kth customer arrived. Since the service times are identically distributed, the waiting time W_k of the kth customer is given by:

$$W_k = R_i u(k) + X_1 + X_2 + \ldots + X_{qk},$$

where $u(k)$ is an indicator function that has a value of 1 if the server was busy when the kth customer arrived and zero otherwise. That is, if N_k defines the total number of customers in the system when the kth customer arrived, then:

$$u(k) = \begin{cases} 1 & N_k > 0 \\ 0 & otherwise. \end{cases}$$

Thus, taking expectations on both sides and noting that N_{qk} and X are independent random variables, and also that $u(k)$ and R_i are independent random variables, we obtain:

$$E[W_k] = E[R_i]E[u(k)] + E[N_{qk}]E[X].$$

Now, from the principles of random incidence in a renewal process, we know that:

$$E[R_i] = \frac{E[X^2]}{2E[X]}.$$

Also,

$$E[u(k)] = 0P[N_k = 0] + 1P[N_k > 0] = P[N_k > 0] = 1 - p_0 = \rho.$$

Finally, from Little's formula, $E[N_{qk}] = \lambda E[W_k]$. Thus, the mean waiting time of the kth customer is given by:

$$E[W_k] = \frac{\rho E[X^2]}{2E[X]} + \lambda E[W_k]E[X] = \frac{\rho E[X^2]}{2E[X]} + \rho E[W_k]$$
$$= \frac{\rho E[X^2]}{2(1-\rho)E[X]} = \frac{\lambda E[X^2]}{2(1-\rho)}.$$

Since the experience of the kth customer is a typical experience, we conclude that the mean waiting time in an M/G/1 queue is given by:

$$E[W] = \frac{\rho E[X^2]}{2(1-\rho)E[X]} = \frac{\lambda E[X^2]}{2(1-\rho)}.$$

Thus, the expected number of customers in the system is given by:

$$E[N] = \rho + E[N_q] = \rho + \frac{\lambda^2 E[X^2]}{2(1-\rho)}.$$

This expression is called the *Pollaczek–Khinchin formula*. It is sometimes written in terms of the coefficient of variation C_X of the service time. The square of C_X is defined as follows:

$$C_X^2 = \frac{\sigma_X^2}{(E[X])^2} = \frac{E[X^2] - (E[X])^2}{(E[X])^2} = \frac{E[X^2]}{(E[X])^2} - 1.$$

Thus, the second moment of the service time becomes $E[X^2] = (1 + C_X^2)(E[X])^2$, and the Pollaczek–Khinchin formula becomes:

$$E[N] = \rho + \frac{\lambda^2 (E[X])^2 (1 + C_X^2)}{2(1-\rho)} = \rho + \frac{\rho^2 (1 + C_X^2)}{2(1-\rho)}.$$

Similarly, the mean waiting time becomes:

$$E[W] = \frac{\lambda E[X^2]}{2(1-\rho)} = \frac{\lambda(1 + C_X^2)(E[X])^2}{2(1-\rho)} = \frac{\rho(1 + C_X^2)E[X]}{2(1-\rho)}.$$

3.7.1 Waiting Time Distribution of the M/G/1 Queue

We can obtain the distribution of the waiting time as follows. Let N_k denote the number of customers left behind by the kth departing customer, and let A_k denote the number of customers that arrive during the service time of the kth customer. Then we obtain the following relationship:

$$N_{k+1} = \begin{cases} N_k - 1 + A_{k+1} & N_k > 0 \\ A_{k+1} & N_k = 0. \end{cases}$$

Thus, we see that $\{N_k\}_{k=0}^{\infty}$ forms a Markov chain called the imbedded M/G/1 Markov chain. Let the transition probabilities of the imbedded Markov chain be defined as follows:

$$p_{ij} = P[N_{k+1} = j | N_k = i].$$

Since N_k cannot be greater than $N_{k+1} + 1$ we have that $p_{ij} = 0$ for all $j < i - 1$. For $j \geq i - 1$, p_{ij} is the probability that exactly $j - i + 1$ customers arrived during the service time of the $(k+1)$th customer, $i > 0$. Also, since the kth customer left the system empty in state 0, p_{0j} represents the probability that exactly j customers arrived while the $(k+1)$th customer was being served. Similarly, since the kth customer left one customer behind, which is the $(k+1)$th customer, in state 1, p_{1j} is the probability that exactly j customers arrived while the $(k+1)$th customer was being served. Thus, $p_{0j} = p_{1j}$ for all j. Let the random variable A_s denote the number of customers that arrive during a service time. Then the probability mass function (PMF) of A_s is given by:

$$p_{A_S}(n) = P[A_S = n] = \int_{x=0}^{\infty} \frac{(\lambda x)^n}{n!} e^{-\lambda x} f_X(x) dx \quad n = 0, 1, \ldots.$$

If we define $\alpha_n = P[A_S = n]$, then the state transition matrix of the imbedded Markov chain is given as follows:

$$P = \begin{bmatrix} \alpha_0 & \alpha_1 & \alpha_2 & \alpha_3 & \cdots & \cdots \\ \alpha_0 & \alpha_1 & \alpha_2 & \alpha_3 & \cdots & \cdots \\ 0 & \alpha_0 & \alpha_1 & \alpha_2 & \cdots & \cdots \\ 0 & 0 & \alpha_0 & \alpha_1 & \cdots & \cdots \\ 0 & 0 & 0 & \alpha_0 & \cdots & \cdots \\ \cdots & \cdots & \cdots & \cdots & \cdots & \cdots \\ \cdots & \cdots & \cdots & \cdots & \cdots & \cdots \end{bmatrix}.$$

The state transition diagram is shown in Figure 3.11.

Observe that the z-transform of the PMF of A_S is given by:

$$\begin{aligned} G_{A_S}(z) &= \sum_{n=0}^{\infty} z^n p_{A_S}(n) = \sum_{n=0}^{\infty} z^n \int_{x=0}^{\infty} \frac{(\lambda x)^n}{n!} e^{-\lambda x} f_X(x) dx \\ &= \int_{x=0}^{\infty} \left\{ \sum_{n=0}^{\infty} \frac{(\lambda x z)^n}{n!} \right\} e^{-\lambda x} f_X(x) dx \\ &= \int_{x=0}^{\infty} e^{\lambda x z} e^{-\lambda x} f_X(x) dx = \int_{x=0}^{\infty} e^{-(\lambda - \lambda z)x} f_X(x) dx \\ &= M_X(\lambda - \lambda z), \end{aligned}$$

where $M_X(s)$ is the s-transform of the PDF of X, the service time. That is, the z-transform of the PMF of A_S is equal to the s-transform of the PDF of X evaluated at the point $s = \lambda - \lambda z$. Let $f_T(t)$ denote the PDF of T, the total time in the system. Let K be a random variable that denotes the number of customers that a tagged customer leaves behind. This is the number of customers that arrived during the total time that the tagged customer was in the system. Thus, the PMF of K is given by:

$$p_K(n) = P[K = n] = \int_{x=0}^{\infty} \frac{(\lambda t)^n}{n!} e^{-\lambda t} f_T(t) dt \quad n = 0, 1, \ldots.$$

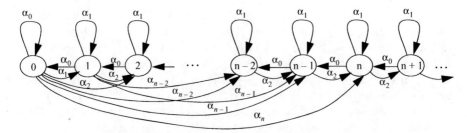

Figure 3.11   Partial state transition diagram for M/G/1 imbedded Markov chain.

As in the case of A_S, it is easy to show that the z-transform of the PMF of K is given by:

$$G_K(z) = M_T(\lambda - \lambda z).$$

Recall that N_k, the number of customers left behind by the kth departing customer, satisfies the relationship:

$$N_{k+1} = \begin{cases} N_k - 1 + A_{k+1} & N_k > 0 \\ A_{k+1} & N_k = 0, \end{cases}$$

where A_k denotes the customers that arrive during the service time of the kth customer. Thus, K is essentially the value of N_k for our tagged customer. It can be shown that the z-transform of the PMF of K is given by:

$$G_K(z) = \frac{(1-\rho)M_X(\lambda - \lambda z)(1-z)}{M_X(\lambda - \lambda z) - z}.$$

Thus, we have that:

$$M_T(\lambda - \lambda z) = \frac{(1-\rho)M_X(\lambda - \lambda z)(1-z)}{M_X(\lambda - \lambda z) - z}.$$

If we set $s = \lambda - \lambda z$, we obtain the following:

$$M_T(s) = \frac{s(1-\rho)M_X(s)}{s - \lambda + \lambda M_X(s)}.$$

This is one of the equations usually called the Pollaczek–Khinchin formula. Finally, since $T = W + X$, which is the sum of two independent random variables, we have that the s-transform of T is given by:

$$M_T(s) = M_W(s)M_X(s).$$

From this we obtain the s-transform of the PDF of W as:

$$M_W(s) = \frac{M_T(s)}{M_X(s)} = \frac{s(1-\rho)}{s - \lambda + \lambda M_X(s)} = \frac{s(1-\rho)}{s - \lambda[1 - M_X(s)]}.$$

This is also called the Pollaczek–Khinchin formula. The expected value is given by:

$$E[W] = -\left[\frac{d}{ds}M_W(s)\right]_{s=0} = \frac{\lambda E[X^2]}{2(1-\rho)},$$

which agrees with the earlier result. Note that the derivation of the previous result was based on an arriving customer's viewpoint while the current result is based on a departing customer's viewpoint. We are able to obtain the same mean waiting for both because the arrival process is a Poisson process. The

Poisson process possesses the so-called PASTA property (Wolff 1982, 1989). PASTA is an acronym for "Poisson arrivals see time averages," which means that customers with Poisson arrivals see the system as if they came into it at a random instant despite the fact that they induce the evolution of the system. Assume that the process is in equilibrium. Let π_k denote the probability that the system is in state k at a random instant, and let π_k' denote the probability that the system is in state k just before a random arrival occurs. Generally, $\pi_k \neq \pi_k'$; however, for a Poisson process, $\pi_k = \pi_k'$.

3.7.2 The M/E$_k$/1 Queue

The M/E$_k$/1 queue is an M/G/1 queue in which the service time has the Erlang-k distribution. It is usually modeled by a process in which service consists of a customer passing, stage by stage, through a series of k independent and identically distributed subservice centers, each of which has an exponentially distributed service time with mean $1/k\mu$. Thus, the total mean service time is $k \times (1/k\mu) = 1/\mu$. The state transition rate diagram for the system is shown in Figure 3.12. Note that the states represent service stages. Thus, when the system is in state 0, an arrival causes it to go to state k; when the system is in state 1, an arrival causes it to enter state $k + 1$, and so on. A completion of service at state j leads to a transition to state $j - 1, j \geq 1$.

While we can analyze the system from scratch, we can also apply the results obtained for the M/G/1 queue. We know for an Erlang-k random variable X, the following results can be obtained:

$$f_X(x) = \frac{(k\mu)^k x^{k-1} e^{-k\mu x}}{(k-1)!},$$

$$M_X(s) = \left(\frac{k\mu}{s + k\mu}\right)^k,$$

$$E[X] = \frac{1}{\mu},$$

$$\sigma_X^2 = \frac{k}{(k\mu)^2} = \frac{1}{k\mu^2},$$

$$C_X^2 = \frac{\sigma_X^2}{(E[X])^2} = \frac{1}{k},$$

$$\rho = \lambda E[X] = \lambda/\mu < 1.$$

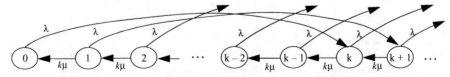

Figure 3.12 State transition rate diagram for the M/E$_k$/1 queue.

Thus, we obtain the following results:

a. The mean waiting time is:

$$E[W] = \frac{\rho(1+C_X^2)E[X]}{2(1-\rho)} = \frac{\rho(k+1)}{2k\mu(1-\rho)}.$$

b. The s-transform of the PDF of the waiting time is:

$$M_W(s) = \frac{s(1-\rho)}{s-\lambda+\lambda M_X(s)} = \frac{s(1-\rho)}{s-\lambda+\lambda\left(\dfrac{k\mu}{s+k\mu}\right)^k}.$$

c. The mean total number of customers in the system is:

$$E[N] = \rho + \frac{\lambda^2(E[X])^2(1+C_X^2)}{2(1-\rho)} = \rho + \frac{\rho^2(k+1)}{2k(1-\rho)}.$$

d. The s-transform of the PDF of the total time in the system is:

$$M_T(s) = \frac{s(1-\rho)M_X(s)}{s-\lambda+\lambda M_X(s)}\Bigg|_{M_X(s)=\left(\frac{k\mu}{s+k\mu}\right)^k}.$$

3.7.3 The M/D/1 Queue

The M/D/1 queue is an M/G/1 queue with a deterministic (or fixed) service time. We can analyze the queueing system by applying the results for M/G/1 queueing system as follows:

$$f_X(x) = \delta(x-1/\mu),$$
$$M_X(s) = e^{-s/\mu},$$
$$E[X] = \frac{1}{\mu},$$
$$\sigma_X^2 = 0,$$
$$C_X^2 = \frac{\sigma_X^2}{(E[X])^2} = 0,$$
$$\rho = \lambda E[X] < 1.$$

Thus, we obtain the following results:

a. The mean waiting time is:

$$E[W] = \frac{\rho(1+C_X^2)E[X]}{2(1-\rho)} = \frac{\rho}{2\mu(1-\rho)}.$$

b. The s-transform of the PDF of the waiting time is:

$$M_W(s) = \frac{s(1-\rho)}{s-\lambda+\lambda M_X(s)} = \frac{s(1-\rho)}{s-\lambda+\lambda e^{-s/\mu}}.$$

c. The mean total number of customers in the system is:

$$E[N] = \rho + \frac{\lambda^2 (E[X])^2 (1+C_X^2)}{2(1-\rho)} = \rho + \frac{\rho^2}{2(1-\rho)}.$$

d. The s-transform of the PDF of the total time in the system is:

$$M_T(s) = \frac{s(1-\rho)e^{-s/\mu}}{s-\lambda+\lambda e^{-s/\mu}}.$$

3.7.4   The M/M/1 Queue

The M/M/1 queue is also an example of the M/G/1 queue with the following parameters:

$$f_X(x) = \mu e^{-\mu x},$$

$$M_X(s) = \frac{\mu}{s+\mu},$$

$$E[X] = \frac{1}{\mu},$$

$$\sigma_X^2 = \frac{1}{\mu^2},$$

$$C_X^2 = \frac{\sigma_X^2}{(E[X])^2} = 1,$$

$$\rho = \lambda E[X] < 1.$$

When we substitute for these parameters in the equations for M/G/1, we obtain the results previously obtained for M/M/1 queueing system.

3.7.5   The M/H$_k$/1 Queue

This is a single-server, infinite-capacity queueing system with Poisson arrivals and hyperexponentially distributed service time of order k. That is, with probability θ_j, an arriving customer will choose to receive service from server j whose service time is exponentially distributed with a mean of $1/\mu_j$, $1 \le j \le k$, where:

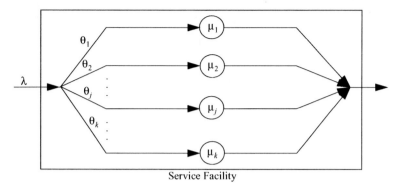

Figure 3.13   A k-stage hyperexponential server.

$$\sum_{j=1}^{k} \theta_j = 1.$$

The system is illustrated in Figure 3.13.
    For this system we have that:

$$f_X(x) = \sum_{j=1}^{k} \theta_j \mu_j e^{-\mu_j x},$$

$$M_X(s) = \sum_{j=1}^{k} \theta_j \left\{ \frac{\mu_j}{s+\mu_j} \right\},$$

$$E[X] = \sum_{j=1}^{k} \left\{ \frac{\theta_j}{\mu_j} \right\},$$

$$E[X^2] = 2 \sum_{j=1}^{k} \frac{\theta_j}{\mu_j^2},$$

$$C_X^2 = \frac{E[X^2] - (E[X])^2}{(E[X])^2} = \frac{E[X^2]}{(E[X])^2} - 1$$

$$= \frac{2 \sum_{j=1}^{k} \frac{\theta_j}{\mu_j^2}}{\left(\sum_{j=1}^{k} \frac{\theta_j}{\mu_j} \right)^2} - 1.$$

For the special case of $k = 2$, that is, for the M/H$_2$/1 queue, we have the following results:

$$f_X(x) = \mu_1 e^{-\mu_1 x} + \mu_2 e^{-\mu_2 x},$$

$$M_X(s) = \frac{\theta_1 \mu_1}{s + \mu_1} + \frac{\theta_2 \mu_2}{s + \mu_2},$$

$$E[X] = \frac{\theta_1}{\mu_1} + \frac{\theta_2}{\mu_2},$$

$$\sigma_X^2 = 2\left(\frac{\theta_1}{\mu_1^2} + \frac{\theta_2}{\mu_2^2}\right),$$

$$C_X^2 = \frac{2(\theta_1 \mu_2^2 - \theta_2 \mu_1^2)}{(\theta_1 \mu_2 + \theta_2 \mu_1)^2}$$

$$\rho = \lambda E[X] < 1.$$

From this we obtain the following performance parameters:

a. The mean waiting time is:

$$E[W] = \frac{\rho(1 + C_X^2)E[X]}{2(1-\rho)} = \frac{\rho\left(\dfrac{\theta_1}{\mu_1} + \dfrac{\theta_2}{\mu_2}\right)\left\{1 + \dfrac{2(\theta_1 \mu_2^2 - \theta_2 \mu_1^2)}{(\theta_1 \mu_2 + \theta_2 \mu_1)^2}\right\}}{2(1-\rho)}.$$

b. The s-transform of the PDF of the waiting time is:

$$M_W(s) = \frac{s(1-\rho)}{s - \lambda + \lambda M_X(s)} = \frac{s(1-\rho)}{s - \lambda + \lambda\left\{\dfrac{\theta_1 \mu_1}{s + \mu_1} + \dfrac{\theta_2 \mu_2}{s + \mu_2}\right\}}.$$

c. The mean total number of customers in the system is:

$$E[N] = \rho + \frac{\lambda^2 (E[X])^2 (1 + C_X^2)}{2(1-\rho)} = \rho + \frac{\rho^2\left\{1 + \dfrac{2(\theta_1 \mu_2^2 - \theta_2 \mu_1^2)}{(\theta_1 \mu_2 + \theta_2 \mu_1)^2}\right\}}{2(1-\rho)}.$$

d. The s-transform of the PDF of the total time in the system is:

$$M_T(s) = \frac{s(1-\rho)\left(\dfrac{\theta_1 \mu_1}{s + \mu_1} + \dfrac{\theta_2 \mu_2}{s + \mu_2}\right)}{s - \lambda + \lambda\left(\dfrac{\theta_1 \mu_1}{s + \mu_1} + \dfrac{\theta_2 \mu_2}{s + \mu_2}\right)}.$$

3.8 PROBLEMS

3.1   People arrive to buy tickets at a movie theater according to a Poisson process with an average rate of 12 customers per hour. The time it takes to complete the sale of a ticket to each person is exponentially distributed with a mean of 3 min. There is only one cashier at the ticket window, and any arriving customer who finds the cashier busy joins a queue that is served in an FCFS manner.

a. What is the probability that an arriving customer does not have to wait?

b. What is the mean number of waiting customers at the window?

c. What is the mean waiting time of an arbitrary customer?

3.2 Cars arrive at a car wash according to a Poisson process with a mean rate of eight cars per hour. The policy at the car wash is that the next car cannot pass through the wash procedure until the car in front of it is completely finished. The car wash has a capacity to hold 10 cars, including the car being washed, and the time it takes a car to go through the wash process is exponentially distributed with a mean of 6 min. What is the average number of cars lost to the car wash company every 10-h day as a result of the capacity limitation?

3.3 A shop has five identical machines that break down independently of each other. The time until a machine breaks down is exponentially distributed with a mean of 10 h. There are two repairmen who fix the machines when they fail. The time to fix a machine when it fails is exponentially distributed with a mean of 3 h, and a failed machine can be repaired by either of the two repairmen. What is the probability that exactly one machine is operational at any one time?

3.4 People arrive at a phone booth according to a Poisson process with a mean rate of five people per hour. The duration of calls made at the phone booth is exponentially distributed with a mean of 4 min.

a. What is the probability that a person arriving at the phone booth will have to wait?

b. The phone company plans to install a second phone at the booth when it is convinced that an arriving customer would expect to wait at least 3 min before using the phone. At what arrival rate will this occur?

3.5 People arrive at a library to borrow books according to a Poisson process with a mean rate of 15 people per hour. There are two attendants at the library, and the time to serve each person by either attendant is exponentially distributed with a mean of 3 min.

a. What is the probability that an arriving person will have to wait?

b. What is the probability that one or both attendants are idle?

3.6 A company is considering how much capacity K to provide in its new service facility. When the facility is completed, customers are expected to arrive at the facility according to a Poisson process with a mean rate of 10 customers per hour, and customers who arrive when the facility is full are lost. The company hopes to hire an attendant to serve at the facility. Because the attendant is to be paid by the hour, the company hopes to get its money's worth by making sure that the attendant is not idle for more than 20% of the time he or she should be working. The service time is expected to be exponentially distributed with a mean of 5.5 min.

a. How much capacity should the facility have to achieve this goal?

b. With the capacity obtained in part a, what is the probability that an arriving customer is lost?

3.7 A small private branch exchange (PBX) serving a startup company can only support five lines for communication with the outside world. Thus, any employee who wants to place an outside call when all five lines are busy is blocked and will have to hang up. A blocked call is considered to be lost because the employee will not make that call again. Calls to the outside world arrive at the PBX according to a Poisson process with an average rate of six calls per hour, and the duration of each call is exponentially distributed with a mean of 4 min.

a. What is the probability that an arriving call is blocked?

b. What is the actual arrival rate of calls to the PBX?

3.8 A cybercafé has six PCs that customers can use for Internet access. These customers arrive according to a Poisson process with an average rate of six per hour. Customers who arrive when all six PCs are being used are blocked and have to go elsewhere for their Internet access. The time that a customer spends using a PC is exponentially distributed with a mean of 8 min.

a. What fraction of arriving customers are blocked?

b. What is the actual arrival rate of customers at the cafe?

c. What fraction of arriving customers would be blocked if one of the PCs is out of service for a very long time?

3.9 Consider a birth and death process representing a multiserver finite population system with the following birth and death rates:

$$\lambda_k = (4 - k)\lambda \quad k = 0, 1, 2, 3, 4$$
$$\mu_k = k\mu \qquad k = 1, 2, 3, 4.$$

a. Find the p_k, $k = 0, 1, 2, 3, 4$ in terms of λ and μ.

b. Find the average number of customers in the system.

3.10 Students arrive at a checkout counter in the college cafeteria according to a Poisson process with an average rate of 15 students per hour. There are three cashiers at the counter, and they provide identical service to students. The time to serve a student by any cashier is exponentially distributed with a mean of 3 min. Students who find all cashiers busy on their arrival join a single queue. What is the probability that at least one cashier is idle?

3.11 Customers arrive at a checkout counter in a grocery store according to a Poisson process with an average rate of 10 customers per hour. There are two clerks at the counter and the time either clerk takes to serve each customer is exponentially distributed with an unspecified mean. If it is desired that the probability that both cashiers are idle is to be no more than 0.4, what will be the mean service time?

3.12 Consider an M/M/1/5 queueing system with mean arrival rate λ and mean service time $1/\mu$ that operates in the following manner. When any customer is in queue, the time until he defects (i.e., leaves the queue without receiving service) is exponentially distributed with a mean of $1/\beta$. It is assumed that when a customer begins receiving service he does not defect.

a. Give the state transition rate diagram of the process.

b. What is the probability that the server is idle?

c. Find the average number of customers in the system.

3.13 Consider an M/M/1 queueing system with mean arrival rate λ and mean service time $1/\mu$ that operates in the following manner. When the number of customers in the system is greater than three, a newly arriving customer joins the queue with probability p and balks (i.e., leaves without joining the queue) with probability $1 - p$.

a. Give the state transition rate diagram of the process.

b. What is the probability that the server is idle?

c. Find the actual arrival rate of customers in the system.

3.14 Consider an M/M/1 queueing system with mean arrival rate λ and mean service time $1/\mu$. The system provides bulk service in the following manner. When the server completes any service, the system returns to the empty state if there are no waiting customers. Customers who arrive while the server is busy join a single queue. At the end of a service completion, all waiting customers enter the service area to begin receiving service.

a. Give the state transition rate diagram of the process.

b. What is the probability that the server is idle?

3.15 Consider an M/G/1 queueing system where service is rendered in the following manner. Before a customer is served, a biased coin whose probability of heads is p is flipped. If it comes up heads, the service time is exponentially distributed with mean $1/\mu_1$. If it comes up tails, the service time is constant at d. Calculate the following:

a. The mean service time, $E[X]$

b. The coefficient of variation, C_X, of the service time

     c. The expected waiting time, $E[W]$

     d. The s-transform of the PDF of W

3.16 Consider a queueing system in which customers arrive according to a Poisson process with rate λ. The time to serve a customer is a third-order Erlang random variable with parameter μ. What is the expected waiting time of a customer?

4

ADVANCED QUEUEING THEORY

4.1 INTRODUCTION

The previous chapter discussed Markovian queueing systems, which are characterized by the fact that either the service times are exponentially distributed or customers arrive at the system according to a Poisson process or both. Specifically, the chapter covered M/M/x and M/G/1 queueing systems. In this chapter we discuss M/G/1 queues with priority as well as a more general queueing system that permits the interarrival and service times to have a general distribution. As in the previous chapter, the emphasis in this chapter is on the steady-state analysis with limited discussion on transient analysis.

4.2 M/G/1 QUEUE WITH PRIORITY

Usually all customers do not have the same urgency. Some customers require immediate attention while others can afford to wait. Thus in many situations, arriving customers are grouped into different priority classes numbered 1 to P such that priority 1 is the highest priority, followed by priority 2, and so on, with priority P being the lowest priority.

There are two main classes of priority queues. These are *preemptive priority* and *nonpreemptive priority*. In a nonpreemptive priority queue, if a higher priority customer arrives while a lower priority customer is being served, the

Fundamentals of Stochastic Networks, First Edition. Oliver C. Ibe.
© 2011 John Wiley & Sons, Inc. Published 2011 by John Wiley & Sons, Inc.

arriving higher priority customer will wait until the lower priority customer's service is completed. Thus, any customer that enters for service will complete the service without interruption. In a preemptive priority queue, if a higher priority customer arrives when a lower priority customer is being served, the arriving customer will preempt the customer being served and begin service immediately. When the preempted customer returns for service, the service can be completed in one of two ways. Under the *preemptive-resume* policy, the customer's service will resume from the point at which it was suspended due to the interruption. Under the *preemptive-repeat* policy, the customer's service will start from the beginning, which means that all the work that the server did prior to the preemption is lost. There are two types of preemptive-repeat priority queueing:*preemptive-repeat without resampling* (also called *preemptive-repeat identical*) and *preemptive-repeat with resampling* (also called *preemptive-repeat different*). In preemptive-repeat identical, the interrupted service must be repeated with an identical requirement, which means that its associated random variable must not be resampled. In preemptive-repeat different, a new service time is drawn from the underlying distribution function.

We assume that customers from priority class p, where $p = 1, 2, \ldots, P$, arrive according to a Poisson process with rate λ_p, and the time to serve a class p customer has a general distribution with mean $\bar{X}_p = E[X_p]$. Define $\rho_p = \lambda_p \bar{X}_p = \lambda_p E[X_p]$ and:

$$\lambda = \sum_{p=1}^{P} \lambda_p,$$

$$E[X] = \bar{X} = \sum_{p=1}^{P} \frac{\lambda_p E[X_p]}{\lambda},$$

$$\rho = \lambda E[X] = \sum_{p=1}^{P} \rho_p < 1,$$

$$\sigma_k = \sum_{p=1}^{k} \rho_p.$$

Thus, ρ_p is the utilization of the server by priority class p customers, λ is the aggregate arrival rate of all customers, and σ_k is the utilization of the server by priority class 1 to k customers. Let W_p be the waiting time of priority class p customers, and let T_p denote the total time a priority class p customer spends in the system.

4.2.1 Nonpreemptive Priority

Consider a tagged priority class 1 customer who arrives for service. Assume that the number of priority class 1 customers waiting when the tagged priority class 1 customer arrived is L_1. The arriving tagged priority class 1 customer has to wait for the customer receiving service to complete his or her service before the priority class 1 customer can receive his or her own service. Assume that the customer receiving service is a priority class p customer whose residual service time is R_p. Thus, the expected waiting time of the tagged customer is given by:

$$E[W_1] = E[L_1]E[X_1] + \sum_{p=1}^{P} \rho_p E[R_p].$$

From Little's formula we have that $E[L_1] = \lambda_1 E[W_1]$. If we define:

$$E[R] = \sum_{p=1}^{P} \rho_p E[R_p],$$

then we have that:

$$E[W_1] = \lambda_1 E[W_1]E[X_1] + E[R] = \rho_1 E[W_1] + E[R].$$

This gives:

$$E[W_1] = \frac{E[R]}{1-\rho_1}.$$

The term $E[R]$ is the expected residual service time. From renewal theory we know that:

$$E[R_p] = \frac{E[X_p^2]}{2E[X_p]} \Rightarrow E[R] = \frac{1}{2}\sum_{p=1}^{P} \frac{\rho_p E[X_p^2]}{E[X_p]} = \frac{1}{2}\sum_{p=1}^{P} \lambda_p E[X_p^2].$$

Thus,

$$E[W_1] = \frac{1}{2(1-\rho_1)}\sum_{p=1}^{P} \lambda_p E[X_p^2].$$

Following the same approach used for a tagged priority class 1 customer, for a priority class 2 customer, we have that:

$$E[W_2] = E[L_1]E[X_1] + E[L_2]E[X_2] + \lambda_1 E[W_2]E[X_1] + \sum_{p=1}^{P} \rho_p E[R_p],$$

where the first term on the right is the mean time to serve the priority class 1 customers who were in queue when the tagged customer arrived, the second term is the mean time to serve the priority class 2 customers who were in queue when the tagged customer arrived, the third term is the mean time to serve those priority class 1 customers who arrived while the tagged priority class 2 customer is waiting to be served, and the fourth term is the mean residual service time of the customer receiving service when the tagged customer arrived. By Little's formula, $E[L_1] = \lambda_1 E[W_1]$ and $E[L_2] = \lambda_2 E[W_2]$. Thus, we have that:

$$E[W_2] = \rho_1 E[W_1] + \rho_2 E[W_2] + \rho_1 E[W_2] + E[R].$$

This gives:

$$E[W_2] = \frac{\rho_1 E[W_1] + E[R]}{1 - \rho_1 - \rho_2} = \frac{E[R]}{(1 - \rho_1)(1 - \rho_1 - \rho_2)}.$$

By continuing in the same way, we can obtain the mean waiting time of a class p customer as:

$$E[W_p] = \frac{E[R]}{(1 - \rho_1 - \ldots - \rho_{p-1})(1 - \rho_1 - \ldots - \rho_p)}$$

$$= \frac{E[R]}{(1 - \sigma_{p-1})(1 - \sigma_p)} = \frac{E[R]}{(1 - \sigma_{p-1})(1 - \rho)}.$$

4.2.2 Preemptive-Resume Priority

In preemptive priority in general, the lower priority class customers are "invisible" to the higher priority class customers. This means that the lower priority class customers do not affect the queues of higher priority class customers. Thus, the mean residual time experienced by class p customers is given by:

$$E[R^p] = \frac{1}{2} \sum_{k=1}^{p} \lambda_k E[X_k^2].$$

Under the preemptive-resume policy, when a customer's service is interrupted by the arrival of a higher priority class customer, the service will be resumed from the point of interruption when all the customers from higher priority classes have been served. The mean service time of a customer is not affected by the interruptions since service always resumes from where it was interrupted. The mean waiting time of a tagged priority class p customer is made up of two parts:

a. The mean time to serve the priority classes $1, \ldots, p$ customers who are ahead of the tagged customer. This component of the mean waiting time is denoted by $E[W_p^a]$. To a priority class p customer, $E[W_p^a]$ is the mean waiting time in an M/G/1 queue without priority when lower priority classes are neglected. Thus,

$$E[W_p^a] = \frac{E[R^p]}{1 - \rho_1 - \ldots - \rho_p},$$

where $E[R^p]$ is the mean residual time experienced by customers of priority class p, as defined earlier.

b. The mean time to serve those customers in priority classes $1, \ldots, p-1$ who arrive while the tagged customer is in queue, which is

$$E[W_p^b] = \sum_{k=1}^{p-1} \lambda_k E[T_p] E[X_k] = \sum_{k=1}^{p-1} \rho_k E[T_p],$$

where $E[T_p] = E[W_p] + E[X_p]$ is the mean time a priority class p customer spends in the system waiting for and receiving service. We refer to T_p as the *sojourn time* of a priority class p customer. Thus, the mean sojourn time of a priority class p customer is given by:

$$E[T_p] = E[X_p] + E[W_p^a] + E[W_p^b]$$

$$= E[X_p] + \frac{E[R^p]}{1 - \rho_1 - \ldots - \rho_p} + E[T_p] \sum_{k=1}^{p-1} \rho_k.$$

From this we obtain:

$$E[T_p] = \frac{E[X_p]}{1 - \rho_1 - \ldots - \rho_{p-1}} + \frac{E[R^p]}{(1 - \rho_1 - \ldots - \rho_p)(1 - \rho_1 - \ldots - \rho_{p-1})}$$

and the mean waiting time is given by:

$$E[W_p] = E[T_p] - E[X_p] = \frac{(\rho_1 + \ldots + \rho_{p-1}) E[X_p]}{1 - \rho_1 - \ldots - \rho_{p-1}}$$

$$= \frac{E[R^p]}{(1 - \rho_1 - \ldots - \rho_{p-1})(1 - \rho_1 - \ldots - \rho_p)}$$

$$= \frac{\sigma_{p-1} E[X_p]}{1 - \sigma_{p-1}} + \frac{E[R^p]}{(1 - \sigma_{p-1})(1 - \sigma_p)}.$$

4.2.3 Preemptive-Repeat Priority

Under the preemptive-resume policy, when a customer's service is interrupted by the arrival of a higher priority class customer, the service will be restarted from the beginning when all the customers from higher priority classes have been served. Let K_p be a random variable that denotes the number of times that a priority class p customer is preempted until his or her service is completed. Let q_p denote the probability that a priority class p customer enters for service and completes the service without being preempted. Then, K_p is a geometrically distributed random variable with probability mass function (PMF) and mean given by:

$$p_{K_p}(k) = q_p(1-q_p)^{k-1} \quad k \geq 1,$$
$$E[K_p] = 1/q_p.$$

The mean waiting time of a tagged priority class p customer is made up of two parts:

a. The mean time to serve the priority classes $1, \ldots, p$ customers who are ahead of the tagged customer. As in the preemptive resume priority case, this component of the waiting time is given by:

$$E[W_p^a] = \frac{E[R^p]}{1-\rho_1-\ldots-\rho_p} = \frac{E[R^p]}{1-\sigma_p}.$$

b. The mean time to serve those customers in priority classes $1, \ldots, p-1$ who arrive while the tagged customer is in queue, which is:

$$E[W_p^b] = \sum_{k=1}^{p-1} \lambda_k E[T_p] E[X_k] = \sum_{k=1}^{p-1} \rho_k E[T_p],$$

where $E[T_p] = E[W_p] + E[Y_p]$ is the mean sojourn time of a priority class p customer, and Y_p is the effective service time, which consists of the actual uninterrupted service time and duration of the $K_p - 1$ times that elapsed before the service was interrupted. Let V_p denote the time that elapses from the instant a priority class p customer commences service until it is preempted. From renewal theory we know that:

$$E[V_p] = E[R_p] = \frac{E[X_p^2]}{2E[X_p]}$$

Thus,

$$E[Y_p] = E[X_p] + \{E[K_p]-1\} E[V_p] = E[X_p] + \frac{\{E[K_p]-1\} E[X_p^2]}{2E[X_p]}$$

$$= E[X_p] + \frac{\{(1/q_p)-1\} E[X_p^2]}{2E[X_p]}.$$

Therefore, the mean sojourn time of a priority class p customer is given by:

$$E[T_p] = E[Y_p] + E[W_p^a] + E[W_p^b]$$

$$= E[X_p] + \frac{\{(1/q_p)-1\} E[X_p^2]}{2E[X_p]} + \frac{E[R^p]}{1-\sigma_p} + E[T_p] \sum_{k=1}^{p-1} \rho_k.$$

This gives:

$$E[T_p] = \frac{E[X_p]}{1-\rho_1-\ldots-\rho_{p-1}} + \frac{(1-q_p)E[X_p^2]}{2q_p(1-\sigma_{p-1})E[X_p]} + \frac{E[R^p]}{(1-\sigma_{p-1})(1-\sigma_p)}.$$

We next compute q_p by noting that it is the probability that no higher priority class customer arrives over the time required to serve a priority class p customer. Since arrivals occur according to a Poisson process, the aggregate arrival rate of priority classes $1, \ldots, p-1$ is given by:

$$\Lambda_{p-1} = \lambda_1 + \ldots + \lambda_{p-1}.$$

Thus,

$$q_p = \int_0^\infty e^{-\Lambda_{p-1}x} f_{X_p}(x)\,dx = M_{X_p}(\Lambda_{p-1}),$$

where $M_{X_p}(s)$ is the s-transform of the probability density function (PDF) of X_p. Finally, the mean waiting time is given by:

$$E[W_p] = E[T_p] - E[X_p]$$
$$= \frac{\sigma_{p-1}E[X_p]}{1-\sigma_{p-1}} + \frac{(1-q_p)E[X_p^2]}{2q_p(1-\sigma_{p-1})E[X_p]} + \frac{E[R^p]}{(1-\sigma_{p-1})(1-\sigma_p)}.$$

4.3 G/M/1 QUEUE

The G/M/1 queue is the dual of the M/G/1 queue. In this system, customers arrive according to a general arrival process with independent and identically distributed interarrival times A with PDF $f_A(t)$ and mean $1/\lambda$. The facility has a single server, and the time X to serve arriving customers is exponentially distributed with mean $1/\mu$. As in the case of the M/G/1 queue, the number $N(t)$ of customers in the system is not Markovian. In this case, the reason is that to completely define a state we need both $N(t)$ and the time that has elapsed since the last arrival. Thus, if Y is the time that has elapsed since the last arrival, then the state of the G/M/1 queue can be defined by the set of pairs $\{(N, Y)\}$, which means that we need a complicated two-dimensional state description. As in the M/G/1 queue, we look for those special points in time where an easier state description can be formulated.

Because of the memoryless nature of the service process, such points that provide an easier state description are those time instants at which customers arrive. At these points, $Y = 0$, which means the state description is captured by N only. Let N_k denote the number of customers that the ith arriving customer sees upon joining the queue, where $N_k = 0, 1, 2, \ldots$. Let T_k denote the time between the kth arrival and the $(k+1)$th arrival. Let S_k denote the number of service completions during T_k, as illustrated in Figure 4.1.

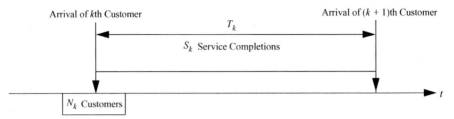

Figure 4.1   Service experience of the kth customer in G/M/1 queue.

Thus, we obtain the following equation:

$$N_{k+1} = \max\left(N_k + 1 - S_k, 0\right) \quad k = 1, 2, \dots.$$

The initial condition is $N_1 = 0$, and we see that the sequence $\{N_k\}_{k=1}^{\infty}$ forms a Markov chain called the G/M/1 *imbedded Markov chain*. Let the transition probabilities be defined by:

$$p_{ij} = P\left[N_{k+1} = j \mid N_k = i\right].$$

It is clear that $p_{ij} = 0$ for all $j > i + 1$. For $j \le i + 1$, p_{ij} represents the probability that exactly $i + 1 - j$ customers are served during the interval between the kth arrival and the $(k + 1)$th arrival, given that the server is busy during this interval. Let r_n denote the probability that n customers are served during an interarrival time. Then r_n is given by:

$$r_n = \int_{t=0}^{\infty} \frac{(\mu t)^n}{n!} e^{-\mu t} f_A(t)\, dt.$$

Thus, the transition probability matrix is given by:

$$P = [p_{ij}] = \begin{bmatrix} 1 - r_0 & r_0 & 0 & 0 & 0 & 0 & \dots \\ 1 - \displaystyle\sum_{k=0}^{1} r_k & r_1 & r_0 & 0 & 0 & 0 & \dots \\ 1 - \displaystyle\sum_{k=0}^{2} r_k & r_2 & r_1 & r_0 & 0 & 0 & \dots \\ 1 - \displaystyle\sum_{k=0}^{3} r_k & r_3 & r_2 & r_1 & r_0 & 0 & \dots \\ \dots & \dots & \dots & \dots & \dots & \dots & \dots \end{bmatrix}.$$

The partial state transition diagram is illustrated in Figure 4.2. In the figure,

$$p_{m0} = 1 - \sum_{k=0}^{m} r_k.$$

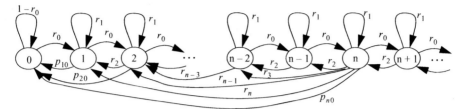

Figure 4.2   Partial state transition diagram for G/M/1 imbedded Markov chain.

Let π_n denote the limiting state probability that the queue is in state n. Then these probabilities must satisfy the following balance equations:

$$\pi_0 = \pi_0 p_{00} + \pi_0 p_{10} + \pi_0 p_{20} + \ldots = \sum_{i=0}^{\infty} \pi_i p_{i0},$$

$$\pi_n = \pi_{n-1} r_0 + \pi_n r_1 + \pi_{n+1} r_2 + \ldots = \sum_{i=0}^{\infty} \pi_{n-1+i} r_i, \ n = 1, 2, \ldots.$$

The solution to this system of equations is of the form:

$$\pi_n = c\beta^n \quad n = 0, 1, \ldots,$$

where c is some constant. Thus, substituting this in the previous equation we obtain:

$$\beta^n = \sum_{i=0}^{\infty} \beta^{n-1+i} r_i \Rightarrow \beta = \sum_{i=0}^{\infty} \beta^i r_i.$$

Since $r_n = \int_{t=0}^{\infty} \dfrac{(\mu t)^n}{n!} e^{-\mu t} f_A(t) \, dt,$ we obtain:

$$\beta = \sum_{i=0}^{\infty} \beta^i \int_{t=0}^{\infty} \frac{(\mu t)^i}{i!} e^{-\mu t} f_A(t) \, dt = \int_{t=0}^{\infty} \left\{ \sum_{i=0}^{\infty} \frac{(\beta \mu t)^i}{i!} \right\} e^{-\mu t} f_A(t) \, dt$$

$$= \int_{t=0}^{\infty} e^{\mu \beta t} e^{-\mu t} f_A(t) \, dt$$

$$= \int_{t=0}^{\infty} e^{-(\mu - \mu \beta)t} f_A(t) \, dt$$

$$= M_A(\mu - \mu \beta),$$

where $M_A(s)$ is the s-transform of the PDF $f_A(t)$ (Ibe 2005). Since we know that $M_A(0) = 1$, $\beta = 1$ is a solution to the functional equation $\beta = M_A(\mu - \mu\beta)$. It can be shown that as long as $\lambda/\mu = \rho < 1$, there is a unique real solution for β in the range $0 < \beta < 1$, which is the solution we are interested in; see Kleinrock (1975). Now, we know that:

$$\sum_{n=0}^{\infty} \pi^n = c \sum_{n=0}^{\infty} \beta^n = 1 = \frac{c}{1-\beta}.$$

Thus, $c = 1 - \beta$, and we obtain:

$$\pi_n = (1-\beta)\beta^n \quad n \geq 0.$$

To find the mean total time in the system (or the sojourn time), let n denote the number of customers in the system when some tagged customer arrived. Since the service time is exponentially distributed, the total time that the tagged customer spends in the system is the time to serve $n + 1$ customers, including the customer, which is given by the following random sum of random variables:

$$T = \sum_{k=1}^{n+1} X_k,$$

where the X_k are independent and identically distributed with the PDF $f_X(x) = \mu e^{-\mu x}, x \geq 0$.

Thus, the s-transform of the PDF of T, $M_T(s)$, can be obtained as follows (see Ibe 2005):

$$T|_{N=n} = X_1 + X_2 + \ldots + X_{n+1}$$

$$M_{T|N=n}(s|n) = \{M_X(s)\}^{n+1}$$

$$M_T(s) = \sum_{n=0}^{\infty} M_{T|N=n}(s|n)\pi_n = \sum_{n=0}^{\infty} \{M_X(s)\}^{n+1} \pi_n$$

$$= \sum_{n=0}^{\infty} \left\{\frac{\mu}{s+\mu}\right\}^{n+1} (1-\beta)\beta^n = \frac{\mu(1-\beta)}{s+\mu} \sum_{n=0}^{\infty} \left\{\frac{\mu\beta}{s+\mu}\right\}^n$$

$$= \frac{\mu(1-\beta)}{s+\mu(1-\beta)}.$$

This means that the sojourn time is exponentially distributed with mean $1/\mu(1 - \beta)$. From this we obtain the following results:

$$E[W] = \frac{1}{\mu(1-\beta)} - \frac{1}{\mu} = \frac{\beta}{\mu(1-\beta)},$$

$$E[N_q] = \lambda E[W] = \frac{\rho\beta}{1-\beta},$$

where $\rho = \lambda/\mu$.

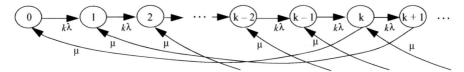

Figure 4.3   State transition rate diagram for the E_k/M/1 queue.

4.3.1   The E_k/M/1 Queue

The E_k/M/1 queue is a G/M/1 queue in which the interarrival time has the Erlang-k distribution. It is usually modeled by a process in which an arrival consists of a customer passing, stage by stage, through a series of k independent and identically distributed substations, each of which has an exponentially distributed service time with mean $1/k\lambda$. Thus, the total mean interarrival time is $k \times (1/k\lambda) = 1/\lambda$. The state transition rate diagram for the system is shown in Figure 4.3. Note that the states represent arrival stages. Thus, when the system is in state 0, an arrival is complete only when the system enters state k and then service can commence. A completion of service causes the system to jump k states to the left.

As an example of a G/M/1 queue, we analyze the system by applying the analysis of the G/M/1 as follows:

$$f_A(t) = \frac{(k\lambda)^k \, t^{k-1} e^{-k\lambda x}}{(k-1)!},$$

$$M_A(s) = \left(\frac{k\lambda}{s+k\lambda}\right)^k,$$

$$\beta = M_A(\mu - \mu\beta) = \left(\frac{k\lambda}{\mu - \mu\beta + k\lambda}\right)^k = \left(\frac{k\rho}{1 - \beta + k\rho}\right)^k.$$

For the special case of $k = 1$, the equation $\beta = M_A(\mu - \mu\beta)$ has the solutions $\beta = 1$ and $\beta = \rho$. Since we seek a solution $0 < \beta < 1$, we accept the solution $\beta = \rho$. Note that when $k = 1$, the system becomes an M/M/1 queue, and the results can be seen to agree with those obtained for the M/M/1 queue in Chapter 3. Similarly, when $k = 2$, the equation becomes:

$$(\beta - 1)\{\beta^2 - \beta(1 + 4\rho) + 4\rho^2\} = 0,$$

whose solutions are:

$$\beta_1 = 1,$$

$$\beta_2 = \frac{1 + 4\rho + \sqrt{1 + 8\rho}}{2} = 2\rho + \frac{1}{2} + \sqrt{2\rho + \frac{1}{4}} \in (1, 4),$$

$$\beta_3 = \frac{1 + 4\rho - \sqrt{1 + 8\rho}}{2} = 2\rho + \frac{1}{2} - \sqrt{2\rho + \frac{1}{4}} \in (0, 1).$$

Since there exist values of $0 \le \rho < 1$ for which β_2 has values greater than 1 and we seek a solution $0 < \beta < 1$, we accept β_3 as the only valid solution. Thus, with this value of β we can obtain all the relevant performance parameters.

4.3.2   The D/M/1 Queue

In this case, the interarrival times are deterministic with a constant value of $1/\lambda$ and service times are exponentially distributed with mean $1/\mu$. There is one server and an infinite capacity. This can be used to model a system with a periodic arrival stream of customers, as is the case with time-division multiplexing (TDM) voice communication systems. Thus, we have that:

$$f_A(t) = \delta(t - 1/\lambda),$$
$$M_A(s) = e^{-s/\lambda},$$
$$\beta = M_A(\mu - \mu\beta) = e^{-(\mu - \mu\beta)/\lambda} = e^{-(1-\beta)/\rho} \Rightarrow \rho \ln \beta = \beta - 1.$$

The last equation can be solved iteratively for a fixed value of ρ to obtain a solution for β in the range $0 < \beta < 1$ that can be used to obtain the performance parameters of the system. That is, for a given value of ρ we can obtain the mean total time in the system, the mean waiting time, and the mean number of customers in queue, respectively, as:

$$E[T] = \frac{1}{\mu(1-\beta)},$$
$$E[W] = \frac{1}{\mu(1-\beta)} - \frac{1}{\mu} = \frac{\beta}{\mu(1-\beta)},$$
$$E[N_q] = \lambda E[W] = \frac{\rho\beta}{1-\beta}.$$

4.3.3   The H$_2$/M/1 Queue

This is a G/M/1 queueing system with the seconnd-order hyperexponential arrival process with rates λ_1 and λ_2. Thus, we have the following parameters:

$$f_A(t) = \theta_1\lambda_1 e^{-\lambda_1 t} + \theta_2\lambda_2 e^{-\lambda_2 t}$$
$$1 = \theta_1 + \theta_2$$
$$E[A] = \frac{1}{\lambda} = \frac{\theta_1}{\lambda_1} + \frac{\theta_2}{\lambda_2} \Rightarrow \lambda = \frac{\lambda_1\lambda_2}{\theta_1\lambda_2 + \theta_2\lambda_1},$$
$$\rho = \lambda/\mu = \frac{\lambda_1\lambda_2}{\mu(\theta_1\lambda_2 + \theta_2\lambda_1)},$$
$$M_A(s) = \frac{\theta_1\lambda_1}{s + \lambda_1} + \frac{\theta_2\lambda_2}{s + \lambda_2},$$
$$\beta = M_A(\mu - \mu\beta) = \frac{\theta_1\lambda_1}{\mu - \mu\beta + \lambda_1} + \frac{\theta_2\lambda_2}{\mu - \mu\beta + \lambda_2}.$$

The last equation implies that:

$$\mu^2\beta^3 - \mu(2\mu + \lambda_1 + \lambda_2)\beta^2 + \{(\mu + \lambda_1)(\mu + \lambda_2) + \mu(\theta_1\lambda_1 + \theta_2\lambda_2)\}\beta$$
$$- \{\theta_1\lambda_1(\mu + \lambda_2) + \theta_2\lambda_2(\mu + \lambda_1)\} = 0.$$

Observe that as discussed earlier, $\beta = 1$ is a solution to the above equation. Thus, we can rewrite the equation as follows:

$$(\beta - 1)\{\mu^2\beta^2 - \mu(\mu + \lambda_1 + \lambda_2)\beta + \lambda_1\lambda_2 + \mu(\theta_1\lambda_1 + \theta_2\lambda_2)\} = 0.$$

Since we exclude the solution $\beta = 1$, we seek the solution to the equation:

$$\mu^2\beta^2 - \mu(\mu + \lambda_1 + \lambda_2)\beta + \lambda_1\lambda_2 + \mu(\theta_1\lambda_1 + \theta_2\lambda_2) = 0,$$

which satisfies the condition $0 < \beta < 1$. Now, the solutions to the above equation are:

$$\beta = \frac{\mu(\mu + \lambda_1 + \lambda_2) \pm \sqrt{\{\mu(\mu + \lambda_1 + \lambda_2)\}^2 - 4\mu^3(\theta_1\lambda_1 + \theta_2\lambda_2)}}{2\mu^2}$$

$$= \frac{(\mu + \lambda_1 + \lambda_2) \pm \sqrt{(\mu + \lambda_1 + \lambda_2)^2 - 4\mu(\theta_1\lambda_1 + \theta_2\lambda_2)}}{2\mu},$$

$$\beta_1 = \frac{\mu + \lambda_1 + \lambda_2}{2\mu} + \sqrt{\left\{\frac{\mu + \lambda_1 + \lambda_2}{2\mu}\right\}^2 - \frac{\theta_1\lambda_1 + \theta_2\lambda_2}{\mu}},$$

$$\beta_2 = \frac{\mu + \lambda_1 + \lambda_2}{2\mu} - \sqrt{\left\{\frac{\mu + \lambda_1 + \lambda_2}{2\mu}\right\}^2 - \frac{\theta_1\lambda_1 + \theta_2\lambda_2}{\mu}}.$$

Thus, we choose β_2 as the right solution. Consider the special case where $\theta_1 = \theta_2 = 0.5$ and $\lambda_1 = 2\lambda_2$, which gives $\rho = 2\lambda_1/3\mu$. The solution becomes:

$$\beta = \frac{1}{2} + \frac{9\rho}{8} - \sqrt{\frac{1}{4} + \frac{81\rho^2}{64}}.$$

With this value of β we can obtain the performance parameters.

4.4   THE G/G/1 QUEUE

We defined a Markovian queueing system as a system in which either the interarrival time or the service time or both are exponentially distributed. Unfortunately many systems in the real world cannot be accurately modeled by the Markovian process. In this section we provide an introduction to the analysis and mean waiting time of the G/G/1 (sometimes known as the GI/G/1, used to emphasize the fact that the interarrival times are statistically independent) queue, which is characterized by the fact that the interarrival time and the service time have a general distribution.

Let $A_1, A_2, \ldots, A_n$, be the arrival times of customers, where A_i is the arrival time of the ith customer. Then $T_n = A_n - A_{n-1}, n \geq 1$ are the interarrival times that are assumed to be independent and identically distributed with a given PDF $f_T(t)$, mean $E[T_n] = 1/\lambda$, and variance σ_T^2, where we assume that $A_0 = 0$. Let $X_1, X_2, \ldots$ denote the service times that are also assumed to be independent and identically distributed with PDF $f_X(x)$, mean $E[X_n] = 1/\mu$, and variance σ_X^2. Interarrival times and service times are assumed to be independent, and service is on a first-come, first-served (FCFS) basis. Let $\rho = \lambda/\mu$.

4.4.1 Lindley's Integral Equation

Let W_n be the waiting time of the nth customer. Then we have that:

$$W_{n+1} = \begin{cases} 0 & \text{if } A_{n+1} \geq A_n + W_n + X_n, \\ A_n + W_n + X_n - A_{n+1} & \text{otherwise,} \end{cases}$$

which follows from Figure 4.4 where C_n denotes the nth customer and we have that $W_{n+1} + T_{n+1} = W_n + X_n$.

Thus, we have that:

$$W_{n+1} = \max[0, A_n - A_{n+1} + W_n + X_n] = \max[0, W_n + X_n - T_{n+1}] = [W_n + X_n - T_{n+1}]^+,$$

where $x^+ = \max[0, x]$. If we define $U_n = X_n - T_{n+1}$, the result becomes:

$$W_{n+1} = [W_n + U_n]^+.$$

Note that the mean and variance of the random variable U_n are given by:

$$E[U_n] = \frac{1}{\mu} - \frac{1}{\lambda},$$

$$\sigma_{U_n}^2 = \sigma_X^2 + \sigma_T^2.$$

We can obtain the cumulative distribution function (CDF) of W_{n+1} as follows. Since W_n and U_n are independent we have that:

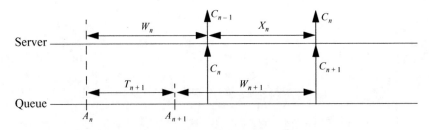

Figure 4.4   Relationships between the queue parameters.

$$F_{W_{n+1}}(w) = P[W_{n+1} \le w] = \begin{cases} P[W_n + U_n \le w] & w \ge 0 \\ 0 & w < 0 \end{cases}$$

$$= P[U_n \le w - W_n] = \int_0^\infty P[U_n \le w - y | W_n = y] dF_{W_n}(y)$$

$$= \int_0^\infty F_{U_n}(w - y) dF_{W_n}(y) \quad w \ge 0,$$

where $F_{U_n}(u)$ is the CDF of $U_n = X_n - T_{n+1}$. That is,

$$F_{U_n}(u) = P[X_n - T_{n+1} \le u] = P[X_n \le u + T_{n+1}]$$

$$= \int_0^\infty P[X_n \le u + v | T_{n+1} = v] dF_{T_{n+1}}(v)$$

$$= \int_0^\infty F_{X_n}(u + v) dF_{T_{n+1}}(v) \quad -\infty < u < \infty,$$

where the last equation follows from the independence of X_n and T_{n+1}. Note that we can also obtain the CDF of W_{n+1} as follows:

$$F_{W_{n+1}}(w) = \int_{-\infty}^w F_{W_n}(w - y) dF_{U_n}(y) \quad w \ge 0.$$

In the limit as $n \to \infty$ we obtain:

$$F_W(w) = \begin{cases} \int_0^\infty F_U(w - y) dF_W(y) & w \ge 0 \\ 0 & w < 0 \end{cases}$$

It can also be shown that

$$F_W(w) = \begin{cases} \int_{-\infty}^w F_W(w - y) dF_U(y) & w \ge 0 \\ 0 & w < 0. \end{cases}$$

These two equations are called the *Lindley's integral equations*. They indicate that the CDF of the waiting time depends only on the difference between the service time and interarrival time distributions rather than on the CDF of the individual random variables.

4.4.2 Laplace Transform of $F_W(w)$

From the equation,

$$F_W(w) = \begin{cases} \int_{-\infty}^w F_W(w - y) dF_U(y) & w \ge 0 \\ 0 & w < 0, \end{cases}$$

we observe that the integral does not have a nonzero numerical value for negative w. We define the integral:

$$F_{W^-}(w) = \begin{cases} \int_{-\infty}^{w} F_W(w-y)dF_U(y) & w < 0 \\ 0 & w \geq 0. \end{cases}$$

Thus, we have that:

$$F_W(w) + F_{W^-}(w) = \int_{-\infty}^{w} F_W(w-y)dF_U(y) \quad -\infty < w < \infty.$$

We define the following two-sided Laplace transforms of $F_W(w)$ and $F_{W^-}(w)$:

$$\psi_W(s) = \int_{-\infty}^{\infty} e^{-sw} F_W(w)dw = \int_{0}^{\infty} e^{-sw} F_W(w)dw,$$

$$\psi_{W^-}(s) = \int_{-\infty}^{\infty} e^{-sw} F_{W^-}(w)dw = \int_{-\infty}^{0} e^{-sw} F_{W^-}(w)dw.$$

We define the Laplace–Stieltjes transform (LST) of the CDF $F_Y(t)$ by:

$$F_Y^*(s) = \int_{0}^{\infty} e^{-st} dF_Y(t).$$

Note that if $F_Y(t)$ is continuous for $t \geq 0$ with $f_Y(t) = dF_Y(t)/dt$, then the LST of $F_Y(t)$ becomes the Laplace transform of $f_Y(t)$.) If $F_T^*(s)$ and $F_X^*(s)$ are the LST of $F_T(t)$ and $F_X(x)$, respectively, then since $U = X - T$ we obtain the LST of $F_U(u)$ as follows:

$$F_U^*(s) = \int_{0}^{\infty} e^{-st} dF_U(u) = F_X^*(s) F_T^*(-s).$$

This follows from the fact that X and T are independent. Recall that:

$$F_W(w) + F_{W^-}(w) = \int_{-\infty}^{w} F_W(w-y)dF_U(y) \quad -\infty < w < \infty,$$

which is a convolution integral. Taking the Laplace transform of the right-hand side gives:

$$\int_{-\infty}^{\infty} \int_{-\infty}^{w} e^{-sw} F_W(w-y)dF_U(y)dw = \int_{-\infty}^{\infty} \int_{-\infty}^{w} e^{-(w-y)s} F_W(w-y)e^{-sw}dF_U(y)dw$$

$$= \left[\int_{-\infty}^{\infty} e^{-(w-y)s} F_W(w-y)dw \right] \left[\int_{-\infty}^{\infty} e^{-sw} dF_U(y) \right]$$

$$= \left[\int_{-\infty}^{\infty} e^{-vs} F_W(v)dv \right] \left[\int_{-\infty}^{\infty} e^{-sw} dF_U(y) \right]$$

$$= \psi_W(s) F_U^*(s),$$

where the second equality is due to the fact that $F_W(w - y) = 0$ when $y \geq w$. Thus we have that:

$$\psi_W(s) + \psi_{W^-}(s) = \psi_W(s) F_U^*(s) = \psi_W(s) F_X^*(s) F_T^*(-s).$$

From this we obtain:

$$\psi_W(s) = \frac{\psi_{W^-}(s)}{F_X^*(s) F_T^*(-s) - 1}.$$

Thus, if the LST of X and T as well as the quantity $\psi_{W^-}(s)$ are known, we obtain the Laplace transform of the CDF of the waiting time. Unfortunately, $\psi_{W^-}(s)$ requires the use of complex number theory, which makes it difficult to solve the Lindley's equation.

4.4.3   Bounds of Mean Waiting Time

A useful upper bound on the mean waiting time is provided by Marshall (1968). A lower bound is given by Marchal (1978). These two bounds state that provided $\rho < 1$, the mean waiting time is bounded as follows:

$$\frac{\rho^2(1 + C_X^2) - 2\rho}{2\lambda(1-\rho)} \leq E[W] \leq \frac{\lambda(\sigma_T^2 + \sigma_X^2)}{2(1-\rho)},$$

where $C_X = \mu \sigma_X$ is the coefficient of variation of the service times.

4.5   SPECIAL QUEUEING SYSTEMS

We conclude this chapter by considering two of the many special queueing systems. Specifically, we consider the following queueing systems:

  a. M/M/1 vacation queueing systems with exceptional first vacation
  b. M/M/1 threshold queueing systems

We have chosen to model these systems as M/M/1 queueing systems for the pedagogical reason of ease of analysis. A more commonly analyzed model is the M/G/1 queueing system.

4.5.1   M/M/1 Vacation Queueing Systems with Exceptional First Vacation

A vacation queueing system is a system in which the server may become unavailable when no customers are waiting at a service completion instant. There are several models of vacation queueing systems, including the single vacation system and the multiple vacation system. In the single vacation queue, a server's vacation begins whenever the system becomes empty. At the end of the vacation, the server returns to begin serving the customers who arrived during his or her vacation, if such customers exist; otherwise, he or she waits until a customer arrives when a busy period commences. The time to serve customers and the duration of a vacation are assumed to be mutually independent random variables. The multiple vacation queue operates in a manner similar to the single vacation queue with the exception that if no customers are found at the end of a vacation, the server immediately commences another vacation. A survey of vacation queueing systems is given in Doshi (1986), and an excellent text on the subject is by Takagi (1991).

We consider a multiple vacation queueing system where customers arrive according to a Poisson process with rate λ. The time to serve a customer is assumed to be exponentially distributed with mean $1/\mu$, where $\mu > \lambda$. We assume that there are two types of vacations: a type 1 vacation that is taken after a busy period of nonzero duration, and a type 2 vacation that is taken when no customers are waiting for the server when he or she returns from a vacation. For ease of analysis we assume that the durations of type 1 vacations are independent of the busy period and are exponentially distributed with mean $1/\gamma_1$. Similarly, durations of type 2 vacations are assumed to be exponentially distributed with mean $1/\gamma_2$.

Let the state of the system be denoted by (r, k), where r is the number of customers in the system, $k = 0$ if the server is actively serving customers, $k = 1$ if the server is on a type 1 vacation, and $k = 2$ if the server is on a type 2 vacation. Thus, the system can be modeled by a continuous-time Markov chain whose state transition rate diagram is shown in Figure 4.5.

Let $p_{n,k}(t)$ denote the probability that the process is in state (n, k) at time t, and let $p_{n,k} = \lim_{t \to \infty} p_{n,k}(t)$ be the limiting value of $p_{n,k}(t)$ and is given by the following theorem:

Theorem 4.1: The steady-state probability $p_{n,k}$ is given by:

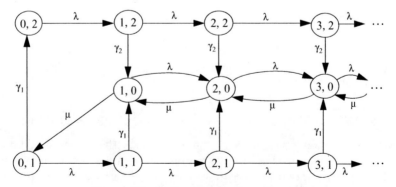

Figure 4.5 State transition rate diagram.

$$p_{n,k} = \begin{cases} \rho\left[\dfrac{\alpha_1\beta_1\left(\beta_1^{n-1}-\rho^{n-1}\right)}{\beta_1-\rho} + \dfrac{\alpha_2\beta_2\left(\beta_2^{n-1}-\rho^{n-1}\right)}{\beta_2-\rho} + \rho^{n-2}\right]p_{1,0} & k=0 \\[4mm] \alpha_1\beta_1^n p_{1,0} & k=1 \\[2mm] \alpha_2\beta_2^n p_{1,0} & k=2 \end{cases},$$

where:

$$p_{1,0} = \frac{(1-\rho)(1-\beta_1)(1-\beta_2)}{(1-\beta_1)(1-\beta_2)+\alpha_1(1-\beta_2)\{1-\rho(1-\beta_1)\}+\alpha_2(1-\beta_1)\{1-\rho(1-\beta_2)\}},$$

$\rho = \lambda/\mu$ is the offered load, $\alpha_1 = \mu/(\lambda+\gamma_1)$, $\alpha_2 = \mu\gamma_1/\{\lambda(\lambda+\gamma_1)\}$, $\beta_1 = \lambda/(\lambda+\gamma_1) < 1$, and $\beta_2 = \lambda/(\lambda+\gamma_2) < 1$.

Proof: From global balance we have that:

$$(\lambda+\gamma_1)p_{0,1} = \mu p_{1,0}.$$

Thus,

$$p_{0,1} = \frac{\mu}{(\lambda+\gamma_1)}p_{1,0} = \alpha_1 p_{1,0},$$

where $\alpha_1 = \mu/(\lambda+\gamma_1)$. Similarly,

$$\lambda p_{0,2} = \gamma_1 p_{0,1},$$

which gives:

$$p_{0,2} = \frac{\gamma_1}{\lambda}p_{0,1} = \left(\frac{\gamma_1}{\lambda}\right)\left(\frac{\mu}{\lambda+\gamma_1}\right)p_{1,0} = \alpha_2 p_{1,0},$$

where $\alpha_2 = \mu\gamma_1/\{\lambda(\lambda+\gamma_1)\}$. Also, for $n = 0, 1, \ldots$, we have that:

$$\lambda p_{n,1} = (\lambda+\gamma_1)p_{n+1,1},$$
$$\lambda p_{n,2} = (\lambda+\gamma_2)p_{n+1,2},$$

which implies that:

$$p_{n+1,1} = \frac{\lambda}{\lambda+\gamma_1}p_{n,1} \quad n = 0, 1, 2, \ldots,$$

$$p_{n+1,2} = \frac{\lambda}{\lambda+\gamma_2}p_{n,2} \quad n = 0, 1, 2, \ldots.$$

Solving the above equations recursively we obtain:

$$p_{n,1} = \left(\frac{\lambda}{\lambda+\gamma_1}\right)^n p_{0,1} = \alpha_1\left(\frac{\lambda}{\lambda+\gamma_1}\right)^n p_{1,0} = \alpha_1\beta_1^n p_{1,0} \quad n = 0, 1, 2, \dots,$$

$$p_{n,2} = \left(\frac{\lambda}{\lambda+\gamma_2}\right)^n p_{0,2} = \alpha_2\left(\frac{\lambda}{\lambda+\gamma_2}\right)^n p_{1,0} = \alpha_2\beta_2^n p_{1,0} \quad n = 0, 1, 2, \dots,$$

where $\beta_1 = \lambda/(\lambda+\gamma_1) < 1$ and $\beta_2 = \lambda/(\lambda+\gamma_2) < 1$. From local balance we obtain:

$$\lambda p_{n,0} + \lambda p_{n,1} + \lambda p_{n,2} = \mu p_{n+1,0} \quad n = 0, 1, 2, \dots.$$

If we define $\rho = \lambda/\mu$, then for $n = 1, 2, \dots$, we obtain:

$$p_{n+1,0} = \rho p_{n,0} + \rho p_{n,1} + \rho p_{n,2}$$
$$= \rho p_{n,0} + \rho\alpha_1\beta_1^n p_{1,0} + \rho\alpha_2\beta_2^n p_{1,0}.$$

Solving recursively we obtain:

$$p_{2,0} = \rho\{\alpha_1\beta_1 + \alpha_2\beta_2 + 1\}p_{1,0}$$
$$= \rho\left\{\frac{\alpha_1\beta_1(\beta_1-\rho)}{\beta_1-\rho} + \frac{\alpha_2\beta_2(\beta_2-\rho)}{\beta_2-\rho} + \rho^0\right\}p_{1,0},$$
$$p_{3,0} = \rho\{\alpha_1\beta_1^2 + \alpha_2\beta_2^2 + \rho\alpha_1\beta_1 + \rho\alpha_2\beta_2 + \rho\}p_{1,0}$$
$$= \rho\{\alpha_1\beta_1(\beta_1+\rho) + \alpha_2\beta_2(\beta_2+\rho) + \rho\}p_{1,0}$$
$$= \rho\left\{\frac{\alpha_1\beta_1(\beta_1^2-\rho^2)}{\beta_1-\rho} + \frac{\alpha_2\beta_2(\beta_2^2-\rho^2)}{\beta_2-\rho} + \rho\right\}p_{1,0},$$
$$p_{4,0} = \rho\{\alpha_1\beta_1^3 + \alpha_2\beta_2^3 + \alpha_1\beta_1^2\rho + \alpha_2\beta_2^2\rho + \alpha_1\beta_1\rho^2 + \alpha_2\beta_2\rho^2 + \rho^2\}p_{1,0}$$
$$= \rho\{\alpha_1\beta_1\{\beta_1^2 + \beta_1\rho + \rho^2\} + \alpha_2\beta_2\{\beta_2^2 + \beta_2\rho + \rho^2\} + \rho^2\}p_{1,0}$$
$$= \rho\left\{\frac{\alpha_1\beta_1(\beta_1^3-\rho^3)}{\beta_1-\rho} + \frac{\alpha_2\beta_2(\beta_2^3-\rho^3)}{\beta_2-\rho} + \rho^2\right\}p_{1,0}.$$

Thus, in general we obtain:

$$p_{n,0} = \rho\left\{\frac{\alpha_1\beta_1(\beta_1^{n-1}-\rho^{n-1})}{\beta_1-\rho} + \frac{\alpha_2\beta_2(\beta_2^{n-1}-\rho^{n-1})}{\beta_2-\rho} + \rho^{n-2}\right\}p_{1,0} \quad n = 1, 2, \dots.$$

From the law of total probability, we have that

$$
1 = \sum_{n=1}^{\infty} p_{n,0} + \sum_{n=0}^{\infty} p_{n,1} + \sum_{n=0}^{\infty} p_{n,2}
$$

$$
= p_{1,0} \left\{ \frac{\alpha_1}{1-\beta_1} + \frac{\alpha_2}{1-\beta_2} + \frac{\alpha_1 \beta_1 \rho}{(1-\rho)(1-\beta_1)} + \frac{\alpha_2 \beta_2 \rho}{(1-\rho)(1-\beta_2)} + \frac{1}{1-\rho} \right\}
$$

$$
= p_{1,0} \left\{ \frac{(1-\beta_1)(1-\beta_2) + \alpha_1(1-\beta_2)\{1-\rho(1-\beta_1)\} + \alpha_2(1-\beta_1)\{1-\rho(1-\beta_2)\}}{(1-\rho)(1-\beta_1)(1-\beta_2)} \right\}.
$$

Thus, we obtain:

$$
p_{1,0} = \frac{(1-\rho)(1-\beta_1)(1-\beta_2)}{(1-\beta_1)(1-\beta_2) + \alpha_1(1-\beta_2)\{1-\rho(1-\beta_1)\} + \alpha_2(1-\beta_1)\{1-\rho(1-\beta_2)\}},
$$

which completes the proof.

The mean number of customers in the system is given by:

$$
E[N] = \sum_{n=1}^{\infty} n p_{n,0} + \sum_{n=0}^{\infty} n p_{n,1} + \sum_{n=0}^{\infty} n p_{n,2}
$$

$$
= \left\{ \frac{\alpha_1 \beta_1 \rho(2-\rho-\beta_1)}{(1-\rho)^2(1-\beta_1)^2} + \frac{\alpha_2 \beta_2 \rho(2-\rho-\beta_2)}{(1-\rho)^2(1-\beta_2)^2} \right.
$$

$$
\left. + \frac{1}{(1-\rho)^2} + \frac{\alpha_1 \beta_1}{(1-\beta_1)^2} + \frac{\alpha_2 \beta_2}{(1-\beta_2)^2} \right\} p_{1,0}.
$$

Finally, from Little's formula, the mean time a customer spends in the system (or mean delay) is given by:

$$
E[T] = E[N]/\lambda.
$$

We assume that $\mu = 1$; thus, $\lambda = \rho$. This implies that:

$$
\alpha_1 = \frac{\mu}{\lambda + \gamma_1} = \frac{1}{\rho(1+\gamma_1/\rho)} = \frac{1}{\rho + \gamma_1},
$$

$$
\alpha_2 = \left(\frac{\gamma_1}{\lambda} \right) \alpha_1 = \frac{\gamma_1}{\rho^2(1+\gamma_1/\rho)} = \frac{\gamma_1}{\rho(\rho+\gamma_1)},
$$

$$
\beta_1 = \frac{\lambda}{\lambda(1+\gamma_1/\lambda)} = \frac{1}{1+\gamma_1/\rho} = \frac{\rho}{\rho+\gamma_1},
$$

$$
\beta_2 = \frac{\lambda}{\lambda(1+\gamma_2/\lambda)} = \frac{1}{1+\gamma_2/\rho} = \frac{\rho}{\rho+\gamma_2}.
$$

We also assume that the mean duration of the type 1 vacation is at least as long as that of the type 2 vacation, which means that $\gamma_2 \geq \gamma_1$. We assume that

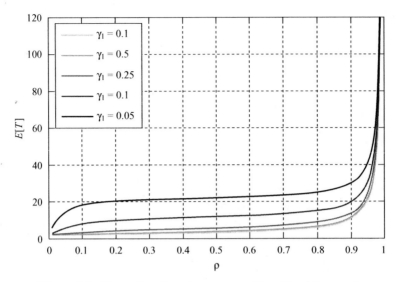

Figure 4.6   Mean time in system versus offered load for $\gamma_2 = 1$.

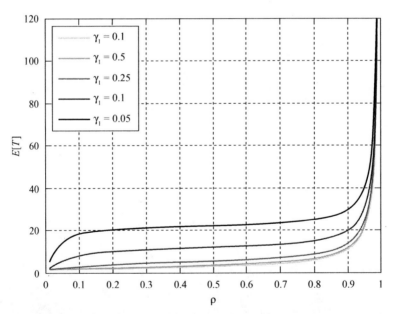

Figure 4.7   Mean time in system versus offered load for $\gamma_2 = 2$.

$\gamma_2 = 1$ and consider different values of $\gamma_1 : \gamma_1 = 1, 0.5, 0.25, 0.1, 0.05$. Figure 4.6 shows the variations of $E[T]$ with ρ. As the figure indicates, $E[T]$ increases as $1/\gamma_1$ increases (corresponding to decreasing γ_1).

Figure 4.7 shows the results for $\gamma_2 = 2$ and $\gamma_1 = 1, 0.5, 0.25, 0.1, 0.05$. The figures show the same trend observed in Figure 4.6.

This queueing system reflects many real-life experiences where some vacations can be used for postprocessing activities while others are actual "breaks" that the server takes.

4.5.2 M/M/1 Threshold Queueing Systems

A multiserver stochastic system can be modeled by a threshold queue if one or more servers are allowed to be idle when the number of customers in the system is below some predefined threshold value. Additional servers are brought in at epochs when the number of customers in the system exceeds the threshold value. Thus, it can also be regarded as a multiserver vacation queueing system in which some of the servers are allowed to take a vacation when the queue length falls below some predefined threshold. They return from vacation when the queue length increases beyond another predefined threshold. The two thresholds are usually different. When the two threshold values are different, the queueing system is said to exhibit a hysteresis behavior. A comprehensive summary of the work on threshold queueing systems with and without hysteresis is given in Ibe and Keilson (1995).

Consider a queueing system with c identical servers. Customers arrive in a Poisson manner with rate λ. The time to serve a customer is exponentially distributed with a mean of $1/\mu$. Assume that the system is empty. Then as customers arrive, they are served by one server, the *primary server*, while the other $c-1$ servers, called the *auxiliary servers*, remain idle. (Note that their idleness is only with respect to serving the arriving customers. In a typical organization these auxiliary servers would be busy with other assignments.) Customers who arrive when the primary server is busy join a queue and are served on an FCFS basis. At the epoch, when the number of customers in the system reaches c, all the auxiliary servers are brought in (or activated) and the system switches to the traditional c-server queueing system. Thus, at the time of the switchover to a c-server queueing system, the queue becomes empty as all the $c-1$ waiting customers will start receiving service. Any other customers who arrive when all the servers are busy will wait and will be served on an FCFS basis. All servers remain active until the system becomes empty again when all the auxiliary servers are retired. They will be brought in again when the number of customers in the system reaches c again. (The auxiliary servers are defined to be active if they have not been retired. While in the active state, an auxiliary server may be temporarily idle if there are no waiting customers and at least one customer but less than c customers is currently receiving service. All the $c-1$ auxiliary servers are retired at the same time, which is when there is no more customer left in the system.)

Let the state of the system be denoted by (r, k), where r is the number of customers in the system, $k = 1$ if only the primary server is active, and $k = 2$ if all the servers are active. The system can then be modeled by a continuous-time Markov chain whose state transition rate diagram is shown in Figure 4.8.

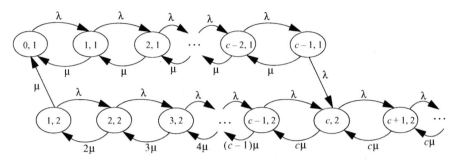

Figure 4.8   State transition rate diagram.

Let $p_{r,k}$ denote the steady-state probability that the process is in state (r, k). Let $E[N]$ and $E[T]$ denote the expected number of customers in the system and the mean time a customer spends in the system, respectively. The value of $p_{r,k}$ is given by the following theorem:

Theorem 4.2: The steady-state probability $p_{r,k}$ is given by:

$$p_{r,k} = \begin{cases} \dfrac{\rho^r(1-\rho^{c-r})}{1-\rho^c}p_{0,1} & k=1, 0 \le r \le c-1 \\[3ex] \dfrac{\rho^c(1-\rho)}{r!(1-\rho^c)}\displaystyle\sum_{j=0}^{r-1} j!\rho^{r-1-j}p_{0,1} & k=2, 1 \le r \le c \\[3ex] \gamma^{r-c}p_{c,2} & k=2, r \ge c, \end{cases}$$

where $\rho = \lambda/\mu$ is the offered load and $\gamma = \rho/c$.

Proof: From local balance we have that:

$$\lambda p_{c-1,1} = \mu p_{1,2},$$
$$p_{r,2} = \gamma^{r-c} p_{c,2} \quad r \ge c.$$

The analysis of state $(r, k), r \le c$, uses the current loop method introduced in Keilson et al. (1968). This method likens the ergodic probability transfer between adjacent states to the current flow across a resistor. As shown in Figure 4.9, the magnitude of this current is:

$$I = \lambda p_{c-1,1} = \mu p_{1,2}.$$

For state $(r, 1), 0 \le r \le c - 1$, the following holds:

$$I = \lambda p_{r-1,1} - \mu p_{r,1}.$$

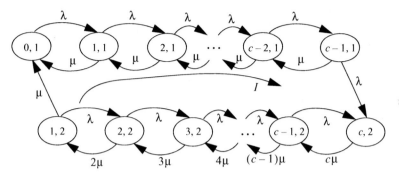

Figure 4.9 The current in the loop.

Substituting for I we obtain:

$$\lambda p_{c-1,1} = \lambda p_{r-1,1} - \mu p_{r,1}.$$

At $r = c - 1$, we obtain:

$$\lambda p_{c-1,1} = \lambda p_{c-2,1} - \mu p_{c-1,1}.$$

From this we obtain:

$$p_{c-2,1} = \left[1 + \frac{1}{\rho}\right] p_{c-1,1} = \left[\frac{1+\rho}{\rho}\right] p_{c-1,1} = \left[\frac{1-\rho^2}{\rho(1-\rho)}\right] p_{c-1,1}.$$

Similarly, at $r = c - 2$ we obtain:

$$\lambda p_{c-1,1} = \lambda p_{c-3,1} - \mu p_{c-2,1}.$$

From this we obtain:

$$p_{c-3,1} = p_{c-1,1} + \frac{1}{\rho} p_{c-2,1} = \left[1 + \frac{1+\rho}{\rho^2}\right] p_{c-1,1} = \left[\frac{1+\rho+\rho^2}{\rho^2}\right] p_{c-1,1}$$

$$= \left[\frac{1-\rho^3}{\rho^2(1-\rho)}\right] p_{c-1,1}.$$

Proceeding in the same manner, it can be shown that:

$$p_{0,1} = \left[\frac{1-\rho^c}{\rho^{c-1}(1-\rho)}\right] p_{c-1,1}.$$

From this we obtain:

$$p_{c-1,1} = \frac{\rho^{c-1}(1-\rho)}{1-\rho^c} p_{0,1}.$$

Substituting this result in the previous results we obtain:

$$p_{r,1} = \left[\frac{\rho^r(1-\rho^{c-r})}{1-\rho^c}\right] p_{0,1} \quad 0 \le r \le c-1.$$

Similarly, for $(r, 2), 1 \le r \le c$, we have that:

$$I = (r+1)\mu p_{r+1,2} - \lambda p_{r,2}.$$

Substituting $I = \mu p_{1,2}$ we obtain:

$$\mu p_{1,2} = (r+1)\mu p_{r+1,2} - \lambda p_{r,2}.$$

At $r = 1$ we obtain:

$$\mu p_{1,2} = 2\mu p_{2,2} - \lambda p_{1,2},$$

from which we obtain:

$$p_{2,2} = \frac{1}{2}[1+\rho] p_{1,2}.$$

At $r = 2$ we obtain:

$$\mu p_{1,2} = 3\mu p_{3,2} - \lambda p_{2,2},$$

which gives:

$$p_{3,2} = \frac{1}{3\mu}\left[\mu + \frac{\lambda}{2}(1+\rho)\right] p_{1,2} = \frac{1}{6}[2+\rho+\rho^2] p_{1,2}.$$

Also, at $r = 3$ we obtain:

$$\mu p_{1,2} = 4\mu p_{4,2} - \lambda p_{3,2},$$

which gives:

$$p_{4,2} = \frac{1}{4\mu}[\mu p_{1,2} + \lambda p_{3,2}] = \frac{1}{4}\left[1 + \frac{\rho}{6}(2 + \rho + \rho^2)\right]p_{1,2} = \frac{1}{24}[6 + 2\rho + \rho^2 + \rho^3]p_{1,2}.$$

From these results it can be shown that:

$$p_{r,2} = \frac{1}{r!}\sum_{j=0}^{r-1} j! \rho^{r-j-1} p_{1,2} \quad 1 \le r \le c.$$

Since $I = \lambda p_{c-1,1} = \mu p_{1,2}$, we have that:

$$p_{1,2} = \rho p_{c-1,1}$$
$$= \rho\left[\frac{\rho^{c-1}(1-\rho)}{1-\rho^c} p_{0,1}\right] = \frac{\rho^c(1-\rho)}{1-\rho^c} p_{0,1}.$$

Thus, we obtain:

$$p_{r,2} = \left\{\frac{\rho^c(1-\rho)}{r!(1-\rho^c)}\sum_{j=0}^{r-1} j! \rho^{r-j-1}\right\}p_{0,1} \quad 1 \le r \le c.$$

Finally, for $r > c$ we use standard M/M/1 queueing analysis techniques to obtain the following:

$$p_{r,2} = \left(\frac{\lambda}{c\mu}\right)^{r-c} p_{c,2} = \gamma^{r-c} p_{c,2}$$
$$= \frac{\gamma^{r-c}\rho^c(1-\rho)}{c!(1-\rho^c)}\sum_{j=0}^{c-1} j! \rho^{c-1-j} p_{0,1} \quad r > c.$$

From the law of total probability we obtain:

$$1 = \sum_{r=0}^{c-1} p_{r,1} + \sum_{r=1}^{c} p_{r,2} + \sum_{r=c+1}^{\infty} p_{r,2}.$$

That is,

$$1 = \left\{\sum_{r=0}^{c-1}\frac{\rho^r(1-\rho^{c-r})}{1-\rho^c} + \sum_{r=1}^{c}\frac{\rho^c(1-\rho)}{r!(1-\rho^c)}\sum_{j=0}^{r-1} j!\rho^{r-j-1} + \sum_{r=c+1}^{\infty}\frac{\gamma^{r-c}\rho^c(1-\rho)}{c!(1-\rho^c)}\sum_{j=0}^{c-1} j!\rho^{c-1-j}\right\}p_{0,1}$$

$$= \left\{\sum_{r=0}^{c-1}\frac{\rho^r(1-\rho^{c-r})}{1-\rho^c} + \sum_{r=1}^{c}\frac{\rho^c(1-\rho)}{r!(1-\rho^c)}\sum_{j=0}^{r-1} j!\rho^{r-j-1} + \frac{\gamma\rho^c(1-\rho)}{c!(1-\rho^c)(1-\gamma)}\sum_{j=0}^{c-1} j!\rho^{c-1-j}\right\}p_{0,1}.$$

From this we obtain $p_{0,1}$. This completes the proof.

The mean number of customers in the system is given by:

$$E[N] = \sum_{r=0}^{c-1} rp_{r,1} + \sum_{r=1}^{c} rp_{r,2} + \sum_{r=c+1}^{\infty} rp_{r,2}.$$

Now,

$$\sum_{r=c+1}^{\infty} rp_{r,2} = \sum_{r=c+1}^{\infty} r\frac{\gamma^{r-c}\rho^{c}(1-\rho)}{c!(1-\rho^{c})}\sum_{j=0}^{c-1} j!\rho^{c-1-j}p_{0,1} = \frac{\rho^{c}(1-\rho)}{c!(1-\rho^{c})}\sum_{j=0}^{c-1} j!\rho^{c-1-j}p_{0,1}\sum_{r=c+1}^{\infty} r\gamma^{r-c}$$

$$= \frac{\rho^{c}(1-\rho)}{c!(1-\rho^{c})}\sum_{j=0}^{c-1} j!\rho^{c-1-j}p_{0,1}\sum_{r=c+1}^{\infty} \gamma\{(r-c-1)+(c+1)\}\gamma^{r-c-1}$$

$$= \frac{\gamma\rho^{c}(1-\rho)}{c!(1-\rho^{c})}\sum_{j=0}^{c-1} j!\rho^{c-1-j}p_{0,1}\left\{\sum_{m=0}^{\infty} m\gamma^{m} + (c+1)\sum_{m=0}^{\infty} \gamma^{m}\right\}$$

$$= \frac{\gamma\rho^{c}(1-\rho)}{c!(1-\rho^{c})}\sum_{j=0}^{c-1} j!\rho^{c-1-j}\left\{\frac{\gamma}{(1-\gamma)^{2}} + \frac{c+1}{1-\gamma}\right\}p_{0,1}.$$

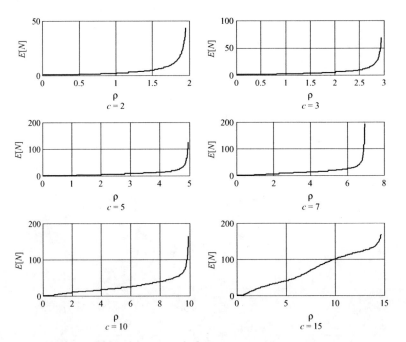

Figure 4.10 Variation of $E[N]$ with ρ for different values of c.

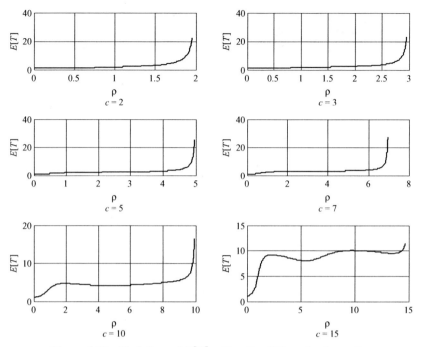

Figure 4.11 Variation of $E[T]$ with ρ for different values of c.

Finally, the mean delay (i.e., the mean time in the system) is obtained from Little's formula as:

$$E[T] = E[N]/\lambda.$$

We assume that $\mu = 1$; thus, $\lambda = \rho$. Figure 4.10 shows the variation of $E[N]$ with ρ for $c = 2, 3, 5, 7, 10, 15$ and Figure 4.11 shows the variation of $E[T]$ with ρ for $c = 2, 3, 5, 7, 10, 15$.

From the figures it can be observed that at a low value of c, the values of both $E[N]$ and $E[T]$ vary with the value of ρ as in a normal queueing system: steadily increasing with ρ and tending to infinity at $\rho = c$. At higher values of c, the values of $E[N]$ tend to show a staircase effect as ρ increases before going to infinity, while the values of $E[T]$ tend to show some ripple effect before tending to infinity. The number of ripples tends to increase as c increases.

4.6 PROBLEMS

4.1 Consider a finite capacity G/M/1 queueing that allows at most three customers in the system including the customer receiving service. The time to serve a customer is exponentially distributed with mean $1/\mu$. As usual, let r_n denote the probability that n customers are served during an interarrival time, $n = 0, 1, 2, 3$, and let $f_A(t)$ be the PDF of the interarrival times. Find r_n in terms of μ and $M_A(s)$, where $M_A(s)$ is the s-transform of $f_A(t)$.

4.2 Consider a queueing system in which the interarrival times of customers are the third-order Erlang random variable with parameter λ. The time to serve a customer is exponentially distributed with mean $1/\mu$. What is the expected waiting time of a customer?

4.3 Consider a G/G/1 queueing system in which the s-transforms of the PDFs of the interarrival times and the service times are given, respectively, by:

$$F_T^*(s) = \frac{2}{(s+1)(s+2)},$$

$$F_X^*(s) = \frac{1}{s+1}.$$

a. Find the mean waiting time.

b. Obtain the upper and lower bounds on the mean waiting time.

4.4 Consider an M/M/2 threshold queueing system in which customers arrive according to a Poisson process with parameter λ and the time to serve a customer is exponentially distributed with mean $1/\mu$ regardless of which server provides the service. Assume that starting with an empty system one server is active until the number of customers in the system reaches $c_1 = 5$ when the second server is activated. When the number of customers in the system decreases to $c_2 = 3$, the server that just completed the last service is deactivated and only one server continues to serve until there is no customer is left in the system. While this one server is active, if the number of customers in the system reaches 5 again, the deactivated server will be brought back to serve.

a. Give the state transition rate diagram of the process.

b. What is the probability that only one server is busy?

c. What is the mean waiting time of customers in the system?

d. Compare the results for the probability that one server is busy and the mean waiting time of customers in the system with the case when $c_1 = c_2 = 5$.

e. Compare the results for the probability that one server is busy and the mean waiting time of customers in the system with the case when $c_1 = c_2 = 3$.

4.5 Consider a finite-buffer, single-server vacation queueing system that operates in the following manner. When the server returns from a vacation, he or she observes the following rule. If there is at least one customer in the system, the server commences service and will serve exhaustively before taking another vacation when the system becomes empty. If there is no customer waiting when the server returns from a vacation, the server will wait for a time that is exponentially distributed with mean $1/\gamma$. If no customer arrives before this waiting time expires, the server commences another vacation. However, if a customer arrives before the waiting time expires, the server will enter a busy period and will serve all customers who arrive during the busy period and then will take a vacation when there is no customer in the system. Assume that customers arrive according to a Poisson process with rate λ, the time to serve a customer is exponentially distributed with mean $1/\mu$, and the duration of a vacation is exponentially distributed with mean $1/\eta$. Also, assume that the maximum capacity of the system is 5. Denote the state of the system by (a, b), where $a = 0$ if the server is on vacation, and $a = 1$ if the server is not on vacation; and b is the number of customers in the system, $b = 0, 1, \ldots, 5$.

a. Give the state transition rate diagram of the process.

b. What is the probability that the server is busy?

c. What is the probability that the server takes a vacation without serving any customer? (Note that this means that the server returns from a vacation and spends the entire waiting period without any customer arriving during the period.)

4.6 Consider an M/M/2 heterogeneous queueing system in which server 1 has a service rate μ_1 and server 2 has a service rate μ_2 such that $\mu_1 > \mu_2$. Customers arrive according to a Poisson process with rate λ. Every arriving customer will try to receive service from server 1 first, but will go to server 2 if server 1 is busy. Let the state of the system be denoted by (ab, c), where $a = 1$ if server 1 is busy and 0 otherwise, $b = 1$ if server 2 is busy and 0 otherwise, and c is the number of customers in the system.

a. Give the state transition rate diagram of the process.

b. What is the probability that both servers are busy?

c. What is the expected number of customers in the system?

4.7 Consider a two-priority queueing system in which priority class 1 (i.e., high-priority) customers arrive according to a Poisson process with rate 2 customers per hour and priority class 2 (i.e., low-priority) customers arrive according to a Poisson process with rate 5 customers per hour. Assume that the time to serve a class 1 customer is exponentially distributed with a mean of 9 minutes, and the time to serve a class 2 customer is constant at 6 minutes. Determine the following:

a. The mean waiting times of classes 1 and 2 customers when nonpreemptive priority is used.

b. The mean waiting times of classes 1 and 2 customers when preemptive resume priority is used.

c. The mean waiting times of classes 1 and 2 customers when preemptive repeat priority is used.

5

QUEUEING NETWORKS

5.1 INTRODUCTION

Many complex systems can be modeled by networks of queues in which customers receive service from one or more servers, where each server has its own queue. Thus, a network of queues is a collection of individual queueing systems that have common customers. In such a system, a departure from one queueing system is often an arrival to another queueing system in the network. Such networks are commonly encountered in communication and manufacturing systems. An interesting example of networks of queues is the set of procedures that patients go through in an outpatient hospital facility where no prior appointments are made before a patient can see the doctor.

In such a facility, patients first arrive at a registration booth where they are first processed on a first-come, first-served (FCFS) basis. Then they proceed to a waiting room to see a doctor on an FCFS basis. After seeing the doctor, a patient may be told to go for a laboratory test, which requires another waiting, or he may be given a prescription to get his medication. If he chooses to fill his prescription from the hospital pharmacy, he joins another queue; otherwise, he leaves the facility and gets the medication elsewhere. Each of these activities is a queueing system that is fed by the same patients. Interestingly, after a laboratory test is completed, a patient may be required to take the result back to the doctor, which means that he rejoins a previous queue. After seeing

Fundamentals of Stochastic Networks, First Edition. Oliver C. Ibe.
© 2011 John Wiley & Sons, Inc. Published 2011 by John Wiley & Sons, Inc.

the doctor a second time, the patient may either leave the facility without visiting the pharmacy, or he will visit the pharmacy and then leave the facility.

A queueing network is completely defined when we specify the external arrival process, the routing of customers among the different queues in the network, the number of servers at each queue, the service time distribution at each queue, and the service discipline at each queue. The network is modeled by a connected directed graph whose nodes represent the queueing systems and the arcs between the nodes have weights that are the *routing probabilities*, where the routing probability from node A to node B is the probability that a customer who leaves the queue represented by node A will next go to the queue represented by node B with the probability labeled on the arc (A, B). If no arc exits between two nodes, then when a customer leaves one node he or she does not go directly to the other node; alternatively, we say that the weight of such an arc is zero.

A network of queues is defined to be either *open* or *closed*. In an open network, there is at least one arc through which customers enter the network and at least one arc through which customers leave the network. Thus, in an open network, a customer cannot be prevented from leaving the network. Figure 5.1 is an illustration of an open network of queues.

By contrast, in a closed network there are no external arrivals and no departures are allowed. In this case, there is a fixed number of customers who are circulating forever among the different nodes. Figure 5.2 is an illustration of a closed network of queues. Alternatively, it can be used to model a finite

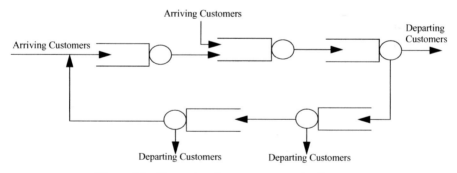

Figure 5.1   Example of an open network of queues.

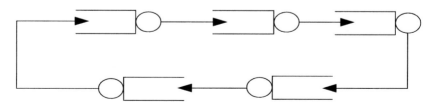

Figure 5.2   Example of a closed network of queues.

capacity system in which a new customer enters the system as soon as one customer leaves the system. One example of a closed network is a computer system that at any time has a fixed number of programs that use the CPU and input/output (I/O) resources.

A network of queues is called a *Markovian network of queues* (or Markovian queueing network) if it can be characterized as follows. First, the service times at the different queues are exponentially distributed with possibly different mean values. Second, if they are open networks of queues, then external arrivals at the network are according to Poisson processes. Third, the transitions between successive nodes (or queues) are independent. Finally, the system reaches a steady state.

To analyze the Markovian network of queues, we first consider a very important result called the *Burke's output theorem* (Burke 1956).

5.2 BURKE'S OUTPUT THEOREM AND TANDEM QUEUES

Burke proved that the departure process of an M/M/c queue is Poisson. Specifically, Burke's theorem states that for an M/M/c queue in the steady state with arrival rate λ, the following results hold:

a. The departure process is a Poisson process with rate λ. Thus, the departure process is statistically identical to the arrival process.
b. At each time t, the number of customers in the system is independent of the sequence of departure times prior to t.

An implication of this theorem is as follows. Consider a queueing network with N queues in tandem, as shown in Figure 5.3. The first queue is an M/M/1 queue through which customers arrive from outside with rate λ. The service rate at queue i is μ_i, $1 \leq i \leq N$, such that $\lambda/\mu_i = \rho_i < 1$. That is, the system is stable. Thus, the other queues are x/M/1 queues and for now we assume that we do not know precisely what "x" is. According to Burke's theorem, the arrival process at the second queue, which is the output process of the first queue, is Poisson with rate λ. Similarly, the arrival process at the third queue, which is the output process of the second queue, is Poisson with rate λ, and so on. Thus, x = M and each queue is essentially an M/M/1 queue.

Assume that customers are served on an FCFS basis at each queue, and let K_i denote the steady-state number of customers at queue i, $1 \leq i \leq N$. Since a departure at queue i results in an arrival at queue $i+1$, $i = 1, 2, \ldots, N-1$, it

Figure 5.3 A network of n queues in tandem.

can be shown that the joint probability of queue lengths in the network is given by:

$$P[K_1 = k_1, K_2 = k_2, \dots, K_N = k_N] = (1-\rho_1)\rho_1^{k_1}(1-\rho_2)\rho_2^{k_2} \dots (1-\rho_N)\rho_N^{k_N}$$

$$= \prod_{i=1}^{N}(1-\rho_i)\rho_i^{k_i}.$$

Since the quantity $(1-\rho_i)\rho_i^{k_i}$ denotes the steady-state probability that the number of customers in an M/M/1 queue with utilization factor ρ_i is k_i, the above result shows that the joint probability distribution of the number of customers in the network of N queues in tandem is the product of the N individual probability distributions. That is, the numbers of customers at distinct queues at a given time are independent. For this reason, the solution is said to be a *product-form solution*, and the network is called a *product-form queueing network*.

A further generalization of this system of tandem queues is a *feed-forward queueing network* where customers can enter the network at any queue but a customer cannot visit a previous queue, as shown in Figure 5.4.

Let γ_i denote the external Poisson arrival rate at queue i and let p_{ij} denote the probability that a customer that has finished receiving service at queue i will next proceed to queue j, where p_{i0} denotes the probability that the customer leaves the network after receiving service at queue i. Then the rate λ_j at which customers arrive at queue j (from both outside and from other queues in the network) is given by:

$$\lambda_1 = \gamma_1$$
$$\lambda_j = \gamma_j + \sum_{i<j} \lambda_i p_{ij} \quad j = 2, \dots, M.$$

It can be shown that if the network is stable (i.e., $\lambda_j/\mu_j = \rho_j < 1$ for all j), then each queue is essentially an M/M/1 queue and a product-form solution exists. That is,

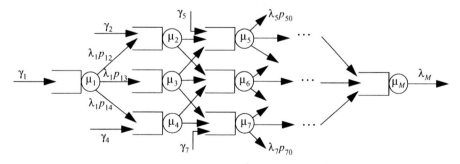

Figure 5.4 A feed-forward queueing network.

$$P[N_1 = n_1, N_2 = n_2, \dots, N_M = n_M] = (1 - \rho_1)\rho_1^{n_1}(1 - \rho_2)\rho_2^{n_2} \dots (1 - \rho_M)\rho_M^{n_M}$$

$$= \prod_{i=1}^{M} (1 - \rho_i)\rho_i^{n_i}.$$

When a customer is allowed to visit a previous queue (i.e., a feedback loop exists), the output process of M/M/1 queue is no longer Poisson because the arrival process and the departure process are no longer independent. For example, if with probability 0.9 a customer returns to the queue after receiving service, then after each service completion there is a very high probability of an arrival (from the same customer that was just served). Markovian queueing networks with feedback are called *Jackson networks*.

5.3 JACKSON OR OPEN QUEUEING NETWORKS

As stated earlier, Jackson networks allow feedback loops, which means that a customer can receive service from the same server multiple times before exiting the network. As previously discussed, because of the feedback loop, the arrival process at each such queue is no longer Poisson. The question becomes this: Is a Jackson network a product-form queueing network?

Consider an open network of N queues in which customers arrive from outside the network to queue i according to a Poisson process with rate γ_i and queue i has c_i identical servers. The time to serve a customer at queue i is exponentially distributed with mean $1/\mu_i$. After receiving service at queue i, a customer proceeds to queue j with probability p_{ij} or leaves the network with probability p_{i0}, where:

$$p_{i0} = 1 - \sum_{j=1}^{N} p_{ij} \quad i = 1, 2, \dots, N.$$

Thus, λ_i, the aggregate arrival rate at queue i, is given by:

$$\lambda_i = \gamma_i + \sum_{j=1}^{N} \gamma_j p_{ji} \quad i = 1, 2, \dots, N.$$

Let $\lambda = [\lambda_1, \lambda_2, \dots, \lambda_N], \gamma = [\gamma_1, \gamma_2, \dots, \gamma_N]$ and let P be the $N \times N$ routing matrix $[p_{ij}]$. Thus, we may write:

$$\lambda = \gamma + \lambda P \Rightarrow \lambda[I - P] = \gamma.$$

Because the network is open and any customer in any queue eventually leaves the network, each entry of the matrix P^n converges to zero as $n \to \infty$. This means that the matrix $I - P$ is invertible and the solution of the preceding equation is:

$$\lambda = \gamma [I - P]^{-1}.$$

We assume that the system is stable, which means that $\rho_i = \lambda_i / c_i \mu_i < 1$ for all i. Let K_i be a random variable that defines the number of customers at queue i in the steady state. We are interested in joint probability mass function $P[K_1 = k_1, K_2 = k_2, \ldots, K_N = k_N]$. Jackson's theorem states that this joint probability mass function is given by the following product-form solution:

$$P[K_1 = k_1, K_2 = k_2, \ldots, K_N = k_N] = P[K_1 = k_1]P[K_2 = k_2]\ldots P[K_N = k_N]$$

$$= \prod_{i=1}^{N} P[K_i = k_i],$$

where $P[K_i = k_i] = p_{K_i}(k_i)$ is the steady-state probability that there are k_i customers in an M/M/c_i queueing system with arrival rate λ_i and service rate $c_i \mu_i$. That is, the theorem states that the network acts as if each queue i is an independent M/M/c_i queueing system. Equivalently, we may write as follows for the case of M/M/1 queueing system:

$$p_K(k) = p_{K_1 K_2 \ldots K_N}(k_1, k_2, \ldots, k_N) = \prod_{i=1}^{N}(1 - \rho_i)\rho_i^{k_i} = \frac{1}{G(N)} \prod_{i=1}^{N} \rho_i^{k_i},$$

$$G(N) = \prod_{i=1}^{N}(1 - \rho_i)^{-1}.$$

Example 5.1: Consider an open Jackson network with $N = 2$, as shown in Figure 5.5. This can be used to model a computer system in which new programs arrive at a CPU according to a Poisson process with rate γ. After receiving service with an exponentially distributed time with a mean of $1/\mu_1$ at the CPU, a program proceeds to the I/O with probability p or leaves the system with probability $1 - p$. At the I/O the program receives an exponentially distributed service with a mean of $1/\mu_2$ and immediately returns to the CPU for further processing. Assume that programs are processed in a first in, first out (FIFO) manner. Obtain the joint PMF of the number of programs at each node.

Solution: The aggregate arrival rates of programs to the two queues are:

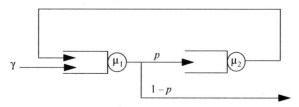

Figure 5.5 An example of an open Jackson network.

$$\lambda_1 = \gamma + \lambda_2,$$
$$\lambda_2 = p\lambda_1.$$

From these two equations we obtain the solutions $\lambda_1 = \gamma/(1 - p)$ and $\lambda_2 = p\gamma/(1 - p)$. Thus, since $c_1 = c_2 = 1$, we have that:

$$P[K_1 = k_1, K_2 = k_2] = P[K_1 = k_1]P[K_1 = k_1] = (1 - \rho_1)\rho_1^{k_1}(1 - \rho_2)\rho_2^{k_2},$$

where $\rho_1 = \lambda_1/\mu_1$ and $\rho_2 = \lambda_2/\mu_2$.

5.4   CLOSED QUEUEING NETWORKS

A *closed network*, which is also called a *Gordon–Newell network* or a *closed Jackson queueing network*, is obtained by setting $\gamma_i = 0$ and $p_{i0} = 0$ for all i. It is assumed that K customers continuously travel through the network and each node i has c_i servers. Gordon and Newell (1967) proved that this type of network also has a product-form solution of the form:

$$P[K_1 = k_1, K_2 = k_2, \ldots, K_N = k_N] = \frac{1}{G_N(K)} \prod_{i=1}^{N} \frac{\rho_i^{k_i}}{d_i(k_i)},$$

where

$$\rho_i = \alpha_i/\mu_i$$

$$d_i(k_i) = \begin{cases} k_i! & k_i \le c_i \\ (c_i!)c_i^{k_i - c_i} & k_i > c_i, \end{cases}$$

$$G_N(K) = \sum_{k_1 + k_2 + \ldots + k_N = K} \prod_{i=1}^{N} \frac{\rho_i^{k_i}}{d_i(k_i)},$$

and $\alpha = (\alpha_1, \alpha_2, \ldots, \alpha_N)$ is any nonzero solution of the equation $\alpha = \alpha P$, where P is the matrix:

$$P = \begin{bmatrix} p_{11} & p_{12} & \cdots & p_{1N} \\ p_{21} & p_{22} & \cdots & p_{2N} \\ \cdots & \cdots & \cdots & \cdots \\ p_{N1} & p_{N2} & \cdots & p_{NN} \end{bmatrix}.$$

Note that because the flow balance equation can be expressed in the form $[I - P]\alpha = 0$, the vector α is the eigenvector of the matrix $I - P$ corresponding to its zero eigenvalue. Thus, the traffic intensities α_i can only be determined to within a multiplicative constant. However, while the choice of α influences the computation of the normalizing constant $G_N(K)$, it does not affect the occupancy probabilities.

$G_N(K)$ depends on the values of N and K. The possible number of states in a closed queueing network with N queues and K customers is:

$$\binom{K+N-1}{K} = \binom{K+N-1}{N-1}.$$

Thus, it is computationally expensive to evaluate the parameter even for a moderate network. An efficient recursive algorithm called the *convolution algorithm* is used to evaluate $G_N(K)$ for the case where the μ_i are constant. Once $G_N(K)$ is evaluated, the relevant performance measures can be obtained. The convolution algorithm and other algorithms for evaluating $G_N(K)$ are discussed later in this chapter.

Example 5.2: Consider the two-queue network shown in Figure 5.6. Obtain the joint probability $P[K_1 = k_1, K_2 = k_2]$ when $K = 3$ and $c_1 = c_2 = 1$.

Solution: Here $N = 2$ and $P = \begin{bmatrix} 1-p & p \\ 1 & 0 \end{bmatrix}$. Thus, if $\alpha = (\alpha_1, \alpha_2)$, then:

$$\alpha = (\alpha_1, \alpha_2) = \alpha P = (\alpha_1, \alpha_2)\begin{bmatrix} 1-p & p \\ 1 & 0 \end{bmatrix} = (\alpha_1(1-p)+\alpha_2, \alpha_1 p).$$

One nonzero solution to the equation is $\alpha_1 = 1$, $\alpha_2 = p$. Thus,

$$\rho_1 = 1/\mu_1$$
$$\rho_2 = p/\mu_2$$
$$d_i(k_i) = 1$$
$$G_2(3) = \sum_{k_1+k_2=3} \rho_1^{k_1}\rho_2^{k_2} = \rho_2^3 + \rho_1\rho_2^2 + \rho_1^2\rho_2 + \rho_1^3.$$

Thus,

$$P[K_1 = k_1, K_2 = k_2] = P[K_1 = k_1, K_2 = 3-k_1]$$
$$= \frac{\rho_1^{k_1}\rho_2^{3-k_1}}{\rho_2^3 + \rho_1\rho_2^2 + \rho_1^2\rho_2 + \rho_1^3} \quad k_1 = 0, 1, 2, 3.$$

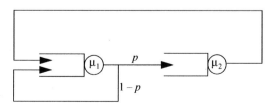

Figure 5.6 An example of a closed Jackson network.

5.5 BCMP NETWORKS

Both the Jackson network and the Gordon–Newell network have a number of limitations. First, they assume that there is a single class of customers at each queue and that service times are exponentially distributed. Second, for open systems, sufficient conditions are provided for the existence of equilibrium distributions, but these conditions are not shown to be necessary. Third, the emphasis is on the product-form nature of the solutions, and customer-oriented measures of system performance, such as the mean waiting time, are largely ignored. Reich (1957) only considers the waiting time of the special case of queues in tandem. Fourth, all customers are assumed to behave in a similar manner in the sense that at each node the service times are identically distributed and service is received on an FCFS basis. Different types of customers are not generally permitted. However, Posner and Bernholtz (1968) considered different customer types in a closed network.

While several network types have been proposed to address these limitations, the most popular among them are the so-called *BCMP networks*. BCMP networks are named after Baskett, Chandy, Muntz, and Palacios who introduced them in Baskett et al. (1975). The BCMP theorem proves a result similar to the product-form result of the Jackson network for a much larger class of queueing networks. In particular, queues can have nonexponentially distributed service times. However when the service time is not exponentially distributed, the queue cannot use the FCFS discipline. Thus, BCMP networks include several classes of customers, different service disciplines, and different interarrival and service time distributions.

5.5.1 Routing Behavior

There are N nodes and $R \geq 1$ classes of customers. For each class the routing probabilities through the network must be specified. A customer belonging to class r can change its class to s after receiving service at a node. Thus, we define the routing matrix $P = [p_{ir,js}]$, where $i, j = 1, \ldots, N$ and $r, s = 1, \ldots, R$. The routing matrix P defines the states by the node-class pair (i, r), where i denotes the index of a node and r denotes the class index. Thus $p_{ir,js}$ is the probability that a customer that completes service at node i as a class r customer will proceed to node j as a class s customer.

A class can be either open in the sense of permitting customers from outside to enter and eventually leave, or closed by not permitting customers to leave. In an open network the arrival process of new customers in the network is Poisson with rate λ. A new arrival is directed to node i and turns to a member of class s with probability $p_{0,is}$. A customer of class r will leave the network after receiving service at node i with probability $p_{ir,0}$.

5.5.2 Service Time Distributions

The exponentially distributed service time in the Jackson network leads to the Markovian property in which the time that a server has already spent on a customer gives no indication of the time remaining until the completion of the service. The BCMP network attempts to conserve this property while allowing more general service time distribution. Recall from Chapter 1 that the Erlang distribution, the hyperexponential distribution, and the Coxian distribution are all methods of stages that permit the service time distribution at each stage to be exponentially distributed. Specifically, let X denote the service time distribution. Then for the kth-order Erlang (or Erlang-k) random variable with parameter μ the probability density function (PDF) is given by:

$$f_{X_k}(x) = \begin{cases} \dfrac{\mu^k x^{k-1} e^{-\mu x}}{(k-1)!} & k = 1, 2, 3, \ldots; x \geq 0 \\ 0 & x < 0. \end{cases}$$

Similarly, for the kth-order hyperexponential random variable H_k that chooses branch i with probability α_i, $i = 1, 2, \ldots, k$, the PDF is given by:

$$f_{H_k}(x) = \sum_{i=1}^{k} \alpha_i \mu_i e^{-\mu_i x}, \quad x \geq 0,$$

$$\sum_{i=1}^{k} \alpha_i = 1.$$

Finally, for the kth-order Coxian random variable C_k, where the mean time spent at stage i is $1/\mu_i$, $i = 1, 2, \ldots, k$, and the customer may choose to receive some service at stage i with probability β_i or leave the system with probability $\alpha_i = 1 - \beta_i$, the s-transform of the PDF of the service time and mean of the service time are given by:

$$M_{C_k}(s) = \sum_{i=1}^{k} L_i \left\{ \prod_{j=1}^{i} \frac{\mu_j}{s + \mu_j} \right\},$$

$$E[C_k] = \sum_{i=1}^{k} L_i \left\{ \sum_{j=1}^{i} \frac{1}{\mu_j} \right\}$$

where L_i is the probability of leaving the system after the ith stage and is given by:

$$L_i = \begin{cases} \alpha_{i+1} \displaystyle\prod_{j=1}^{i} \beta_j & i = 1, 2, \ldots, k-1 \\ \displaystyle\prod_{j=1}^{k} \beta_j & i = k. \end{cases}$$

5.5.3 Service Disciplines

The queues in the BCMP network can use any of the following four disciplines:

1. FCFS at a single-server node where the service time is exponentially distributed with the same mean for all classes of customers. The service rate at node i may depend on the local queue length and is denoted by $\mu_i(k)$ if there are k customers at the node.
2. Processor sharing (PS), which means that a customer at node i receives $1/k$ seconds of service per second in turn if there are $k > 0$ customers at the node. The service time distribution can be a distinct Coxian distribution for each class of customers.
3. Infinite server (IS), which means that the number of servers at a station is sufficiently large to ensure that a new customer arriving at the node starts service immediately without having to wait. The service time distribution can be a distinct Coxian distribution for each class of customers
4. Last-come, first-served preemptive resume (LCFS PR), which means that a newly arriving customer at a node preempts an ongoing service and enters for service. There is a single server at the node; a preempted customer stays at the head of the queue and his or her service is resumed from where it was interrupted when the customer who preempted it completes his or her own service. The service time distribution is a distinct Coaxian distribution for each class of customers.

The service times and routing behavior are independent of each class and of the past. Note that apart from type 1 that operates on the FCFS basis, other models have the property that a customer commences service immediately upon arrival. When service cannot commence immediately upon the customer's arrival, the service time must be exponentially distributed.

5.5.4 The BCMP Theorem

There are two kinds of arrival processes in an open network, which are as follows:

- The arrival process is Poisson where all customers arrive at the network with an overall arrival rate λ that can depend on the number of customers in the network. Thus, if K is the number of customers in the network, we designate the arrival rate $\lambda(K)$. An arriving customer is assigned class r and sent to node i with probability $p_{0,\,ir}$, where:

$$\sum_{i=1}^{N}\sum_{r=1}^{R} p_{0,ir} = 1.$$

- The arrival process consists of M independent Poisson arrival streams. Let $\lambda_i(K_i)$ be the arrival rate of stream i when there are K_i customers of the stream in the network. Each stream constitutes a routing chain, and for each stream m an arrival joins node i with probability $p_{0,\,im}$. A stream is made up of a set of classes, which have the property that a customer from one class in the stream can become a member of another class in the stream. Thus, the number of streams is no greater than the number of classes.

For routing chain (or arrival stream) c, *the traffic equation*, which is the net arrival of class r to node i is:

$$\lambda_{ir} = \lambda_{ir}^0 + \sum_{(j,s)} \lambda_{js} p_{js,ir},$$

where λ_{ir}^0 is the arrival rate of customers from outside the network, which is zero for a closed network. For an open network $\lambda_{ir}^0 = \lambda p_{0,ir}$ if we have one arrival stream, otherwise $\lambda_{ir}^0 = \lambda_r p_{0,ir}$ arrivals per stream per class. For a closed network λ_{ir} is the *visit ratio* of node i for chain i. Consider the following set of equations for each arrival stream:

$$e_{ir} = p_{0,ir} + \sum_{s=1}^{R} \sum_{j=1}^{N} e_{js} p_{js,ir} \quad 1 \le i \le N, 1 \le r \le R.$$

The parameter e_{ir} gives the relative frequency of the number of visits to node i by a customer of class r. For a single-class network we have that:

$$e_i = p_{0,i} + \sum_{j=1}^{N} e_j p_{j,i} \quad 1 \le i \le N.$$

For a closed network, $p_{0,\,i} = 0$ and e_i is not the absolute arrival rate of customers to node i. Also, the ratio λ_i/μ_i is not the actual utilization factor of node i. Let $X_i = (X_{i1}, X_{i2}, \ldots, X_{iR})$ be a vector that denotes the state of node i that depends on the service discipline at the node, and let the state of the network be the vector $X = (X_1, X_2, \ldots, X_N)$.

- For a type 1 node with FCFS service discipline, $X_i = (x_{i1}, x_{i2}, \ldots, x_{ik_i})$, where x_{ij} is the class of the jth customer waiting at node i. The first customer is being served while the other customers are waiting to receive service in the order they came.
- For type 2 or type 3 node with either PS or IS,

$$X_i = ((x_{i1}, s_{i1}), (x_{i2}, s_{i2}), \ldots, (x_{ix_i}, s_{ix_i})),$$

where x_{ij} is the class of the jth customer at node i in the order of arrival and s_{ij} is the stage of the Coxian model.

- For a type 4 node with last-come, first-served (LCFS) service discipline, $X_i = ((x_{i1}, s_{i1}), (x_{i2}, s_{i2}), \ldots, (x_{ix}, s_{ix}))$, where x_{ij} is the class of the jth customer waiting at node i in LCFS order and s_{ij} is the stage of the Coxian model.

Let k_{irm} denote the number of customers of class r at the mth stage at node i. Also, let K_i be the total number of customers at node i, and let K denote the number of customer in the network; that is,

$$K_i = \sum_{r=1}^{R} \sum_{m=1}^{l_{ir}} k_{irm},$$

$$K = \sum_{i=1}^{N} K_i,$$

where l_{ir} is the number of Coaxian stages that class r customers go through at node i. The BCMP theorem states that the steady-state probability distribution in a BCMP network with K customers in the network has the following product form:

$$P[K_1 = k_1, K_2 = k_2, \ldots, K_N = k_N] = \frac{1}{G_N(K)} d(K) \prod_{i=1}^{N} f_i(k_i),$$

where $G_N(K)$ is a normalizing constant chosen to make the equilibrium state probabilities sum to 1, $d(K)$ is a function of the external arrival process only, and the functions $f_i(k_i)$ are the "per-node" steady-state distributions that depend on the type of service at node i. $d(K)$ is given by:

$$d(K) = \begin{cases} \prod_{m=0}^{K-1} \lambda(m) & \text{if the network is open with single arrival stream,} \\ \prod_{m=1}^{M} \prod_{i=0}^{K-1} \lambda_m(i) & \text{if the network is open with multiple arrival streams,} \\ 1 & \text{if the network is closed.} \end{cases}$$

Let L_{irm} denote the probability that a class r customer reaches the mth Coxian stage at node i. Thus, from Chapter 1 we have that $L_{irm} = \beta_1 \beta_2 \ldots, \beta_m$; see Figure 5.7.

Finally, let $1/\mu_{irm}$ denote the mean service time of class r customers at the mth stage of node i. The following results are the values of $f_i(k_i)$ for the different service disciplines:

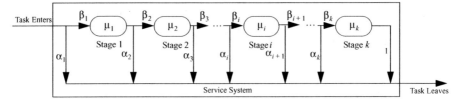

Figure 5.7 Graphical representation of C_k random variable.

- When node i is of the type FCFS (or type 1) we have:

$$f_i(k_i) = \prod_{j=1}^{k_i} \frac{e_{ik_{ij}}}{\mu_i(j)}.$$

- When node i is of the type PS with Coxian distribution (or type 2) we have:

$$f_i(k_i) = k_i! \prod_{r=1}^{R} \prod_{m=1}^{l_{ir}} \frac{e_{ir}^{k_{irm}}}{k_{irm}!} \left(\frac{L_{irm}}{\mu_{irm}}\right)^{k_{irm}}.$$

- When node i is of the type IS with Coxian distribution (or type 3) we have:

$$f_i(k_i) = \prod_{r=1}^{R} \prod_{m=1}^{l_{ir}} \frac{e_{ir}^{k_{irm}}}{k_{irm}!} \left(\frac{L_{irm}}{\mu_{irm}}\right)^{k_{irm}}.$$

- When node is of the type LCFS with Coxian distribution (or type 4) we have:

$$f_i(k_i) = \prod_{j=1}^{k_i} e_{ik_{ij}} \left\{\frac{L_{ik_{ij}s_{ij}}}{\mu_{ik_{ij}s_{ij}}}\right\}.$$

The normalizing constant is given by:

$$G_N(K) = \sum_{(k_i, k_2, \ldots, k_N) \in F} d(K) \prod_{i=1}^{N} f_i(k_i),$$

where F is the set of possible states.

Note that the product-form solution holds under two conditions. The first is that service times are exponentially distributed and customers are served in an FCFS manner. The second is that if service times are not exponentially distributed, then an arriving customer starts receiving service immediately upon its arrival. This is the case in LCFS, IS, and PS service disciplines where an arriving customer does not wait; his or her service commences immediately upon his or her arrival.

5.6 ALGORITHMS FOR PRODUCT-FORM QUEUEING NETWORKS

The product-form queueing networks have solutions that require the evaluation of the function $G_N(K)$. In this section we consider two methods that have been proposed for computing this function efficiently. These are the *convolution algorithm* and the *mean value analysis* (MVA).

5.6.1 The Convolution Algorithm

We first consider a closed queueing network with a single class of customers. We assume that there are K customers and N nodes, where each node has a single server. If K_i denotes the number of customers at node i, then we have that:

$$P[K_1 = k_1, K_2 = k_2, \ldots, K_N = k_N] = \frac{1}{G_N(K)} \prod_{i=1}^{N} \rho_i^{k_i},$$

$$G_N(K) = \sum_{k_1+k_2+\ldots+k_N=K} \prod_{i=1}^{N} \rho_i^{k_i},$$

where $\rho_i = e_i/\mu_i$. Consider the function:

$$g(k, n) = \sum_{k_1+k_2+\ldots+k_N=K} \prod_{i=1}^{n} \rho_i^{k_i},$$

which means that $g(K, N) = G_N(K)$ and $g(k, N) = G_N(k)$ for $k = 0, 1, \ldots, K$. For $k > 0$ and $n > 1$ we have that:

$$g(k, n) = \sum_{k_1+k_2+\ldots+k_{n-1}=k, k_n=0} \prod_{i=1}^{n} \rho_i^{k_i} + \sum_{k_1+k_2+\ldots+k_n=k, k_n>0} \prod_{i=1}^{n} \rho_i^{k_i},$$

$$= g(k, n-1) + \rho_n g(k-1, n)$$

$$g(k, 1) = \rho_1^k \quad \text{for } k = 0, 1, \ldots, K,$$

$$g(0, n) = 1 \quad \text{for } n = 1, 2, \ldots, N,$$

$$g(k, 0) = 0 \quad \text{for } k = 1, 2, \ldots, K.$$

Thus, we can now compute $g(K, N) = G_N(K)$ recursively. The recurrence equation,

$$g(k, n) = g(k, n-1) + \rho_n g(k-1, n),$$

along with the initial conditions $g(0, n) = 1$ and $g(k, 0) = 0$, is the *convolution algorithm*, which was developed by Buzen (1973).

Other performance measures can also be computed. For example, the queue lengths can be obtained as follows. Define:

$$S(K, N) = \left\{ (k_1, k_2, \ldots, k_N) \left| \sum_{i=1}^{N} k_i = K, \quad k_i \geq 0 \; \forall i \right. \right\}.$$

We are interested in finding the probability mass function (PMF) of K_i, $p_{K_i}(k) = P[K_i = k]$. We observe that:

$$P[K_i \geq k] = \sum_{S(K,N) \cap \{K_i \geq k\}} P[K_1 = k_1, K_2 = k_2, \ldots, K_N = k_N]$$

$$= \sum_{S(K,N) \cap \{K_i \geq k\}} \frac{1}{G_N(K)} \prod_{i=1}^{N} \rho_i^{k_i} = \frac{\rho_i^k}{G_N(K)} \sum_{S(K-k,N)} \prod_{j=1}^{N} \rho_j^{k_j}$$

$$= \frac{\rho_i^k G_N(K-k)}{G_N(K)}, \quad k \geq 1.$$

We know that $p_{K_i}(k) = P[K_i = k] = P[K_i \geq k] - P[K_i \geq k+1]$, which means that:

$$p_{K_i}(k) = \frac{\rho_i^k G_N(K-k)}{G_N(K)} - \frac{\rho_i^{k+1} G_N(K-k-1)}{G_N(K)}$$

$$= \frac{\rho_i^k}{G_N(K)} \{ G_N(K-k) - \rho_i G_N(K-k-1) \}.$$

Thus,

$$E[K_i] = \sum_{k=1}^{K} P[K_i \geq k] = \sum_{k=1}^{K} \frac{\rho_i^k G_N(K-k)}{G_N(K)}.$$

Hence once the values of $G_N(k)$, $k = 1, \ldots, K$, are known we can obtain the expected values of the numbers of customers at each node.

The utilization of node i is the probability that there is at least one customer at the node, which is given by:

$$u_i = P[K_i \geq 1] = \frac{\rho_i G_N(K-1)}{G_N(K)}.$$

Finally, if we denote the mean service time at node i by $1/\mu_i$, then the throughput of node i is given by $\gamma_i = u_i \mu_i$. That is,

$$\gamma_i = u_i \mu_i = \frac{e_i G_N(K-1)}{G_N(K)},$$

where e_i is the visit ratio of node i. Using Little's formula we can obtain the mean delay at node i (queueing plus service), $E[T_i]$, which is given by:

$$E[T_i] = \frac{E[K_i]}{\gamma_i} = \frac{\sum_{k=1}^{K} \rho_i^k G_N(K-k)}{e_i G_N(K-1)}.$$

Recall that the e_i are obtained from the system of equations:

$$e_i = \sum_{j=1}^{N} e_j p_{ji} \quad i = 1, \ldots, N.$$

Example 5.3: Consider the network in Figure 5.8. We want to use the convolution method to obtain the mean delay at nodes 1 and 2 with $\mu_1 = \mu_2 = 1$ and $K = 2$.

Solution: We first solve the traffic flow equations for a closed queueing network, which are:

$$e_i = \sum_{j=1}^{N} e_j p_{ji},$$

$$e_1 = \tfrac{1}{2} e_1 + e_2 \implies \tfrac{1}{2} e_1 = e_2.$$

If we fix $e_1 = 1$, then we have that $e_2 = 1/2$. Thus $\rho_1 = e_1/\mu_1 = 1$ and $\rho_2 = e_2/\mu_2 = 1/2$. The recursive equation and the boundary conditions are as follows:

$$g(k,n) = g(k,n-1) + \rho_n g(k-1,n),$$
$$g(k,1) = \rho_1^k \quad \text{for } k = 0, 1, \ldots, K,$$
$$g(0,n) = 1 \quad \text{for } n = 1, 2, \ldots, N,$$
$$g(k,0) = 0 \quad \text{for } k = 1, 2, \ldots, K.$$

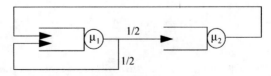

Figure 5.8 Figure for Example 5.3.

Thus, we have that:

$$g(1, 1) = g(1, 0) + \rho_1 g(0, 1) = \rho_1,$$
$$g(1, 2) = g(1, 1) + \rho_2 g(0, 2) = \rho_1 + \rho_2,$$
$$g(2, 1) = \rho_1^2,$$
$$g(2, 2) = g(2, 1) + \rho_2 g(1, 2) = \rho_1^2 + \rho_2\{\rho_1 + \rho_2\} = \rho_1^2 + \rho_2^2 + \rho_1\rho_2.$$

Thus, $G_2(2) = g(2, 2) = \rho_1^2 + \rho_2^2 + \rho_1\rho_2$. From these results we obtain the throughputs as follows:

$$\gamma_i = \frac{e_i G_N(K-1)}{G_N(K)} = \frac{e_i g(1, 2)}{g(2, 2)} = \frac{e_i(\rho_1 + \rho_2)}{\rho_1^2 + \rho_2^2 + \rho_1\rho_2}$$

$$= \begin{cases} 6/7 & i = 1 \\ 3/7 & i = 2. \end{cases}$$

Also, the mean delays are given by:

$$E[T_1] = \frac{\sum\limits_{k=1}^{2} \rho_1^k G_2(2-k)}{e_1 G_2(2-1)} = \frac{\rho_1 G_2(1) + \rho_1^2 G_2(0)}{e_1 G_2(1)} = \frac{\rho_1 g(1, 2) + \rho_1^2 g(0, 2)}{e_1 g(1, 2)}$$

$$= \frac{\rho_1\{\rho_1 + \rho_2\} + \rho_1^2}{e_1\{\rho_1 + \rho_2\}} = \frac{\rho_1\{2\rho_1 + \rho_2\}}{e_1\{\rho_1 + \rho_2\}} = \frac{2\rho_1 + \rho_2}{\mu_1\{\rho_1 + \rho_2\}} = \frac{5}{3},$$

$$E[T_2] = \frac{\sum\limits_{k=1}^{2} \rho_2^k G_2(2-k)}{e_2 G_2(K-1)} = \frac{\rho_2 G_2(1) + \rho_2^2 G_2(0)}{e_2 G_2(1)} = \frac{\rho_2 g(1, 2) + \rho_2^2 g(0, 2)}{e_2 g(1, 2)}$$

$$= \frac{\rho_2\{\rho_1 + \rho_2\} + \rho_2^2}{e_2\{\rho_1 + \rho_2\}} = \frac{\rho_2\{\rho_1 + 2\rho_2\}}{e_2\{\rho_1 + \rho_2\}} = \frac{\rho_1 + 2\rho_2}{\mu_2\{\rho_1 + \rho_2\}} = \frac{4}{3}.$$

Thus, the mean number of customers at each node is given by:

$$E[K_1] = \gamma_1 E[T_1] = \frac{6}{7} \times \frac{5}{3} = \frac{10}{7},$$

$$E[K_2] = \gamma_2 E[T_2] = \frac{3}{7} \times \frac{4}{3} = \frac{4}{7}.$$

For a multiclass network, the convolution algorithm for the case when node is load independent is given by:

$$g(k, n) = g(k, n-1) + \sum_{r=1}^{R} \rho_{nr} g(k - e_r, n),$$

where $k - e_r$ means a network with one customer less in the rth chain. The mean queue length, utilization, and throughput of every node can be calculated as in the single-class case.

5.6.2 The MVA

When only the mean delay is desired, it is possible to compute these means directly through an iterative technique called the MVA. MVA was developed by Reiser and Lavenberg (1980) and is used to solve closed queueing networks with exponential service times. It calculates the expected number of customers at node i, $E[K_i]$, directly without first computing $G_N(K)$ or deriving the stationary distribution in the network. The algorithm is based on the so-called *arrival theorem*, which states that the state of the system at the time of an arrival has a probability distribution equal to that of the steady-state distribution of the network with one customer less. In other words, if we single out a customer as an arrival, he or she should "see" the system without himself or herself in equilibrium.

According to the arrival theorem, when a customer arrives at node i, the expected number of customers that he or she sees at node i is $E[N_i(K-1)]$, where $N_i(K)$ is the number of customers at node i when the total number of customers in the network is K. Thus, assuming that the node uses FCFS and that the mean service time at the node for a single-class system is $1/\mu_i$, the mean delay time at the node is given by:

$$E[T_i(K)] = \frac{1}{\mu_i}\{1 + E[N_i(K-1)]\} \quad i = 1,\ldots,N,$$

since he or she will wait for the $E[N_i(K-1)]$ customers in the system to be served and then receives his or her own service. Let $\lambda(K)$ denote the system throughput and $\lambda_i(K) = \pi_i \lambda(K)$ denote the average customer arrival rate at node i, where π_i are the flows, which are solutions to the system of equations:

$$\pi_i = \sum_{j=1}^{N} \pi_j p_{ji}, \quad \sum_{j=1}^{N} \pi_j = 1.$$

Then from Little's formula, the mean number of customers at node i is given by:

$$E[N_i(K)] = \lambda_i(K)E[T_i(K)] = \lambda(K)\pi_i E[T_i(K)] \quad i = 1,\ldots,N.$$

We also know from Little's law that:

$$K = \lambda(K)\sum_{i=1}^{N} \pi_i E[T_i(K)] \Rightarrow \lambda(K) = K\left[\sum_{i=1}^{N} \pi_i E[T_i(K)]\right]^{-1}.$$

Thus, the mean delay can be solved iteratively using the above three equations by starting with $E[N_i(0)] = 0$.

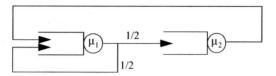

Figure 5.9 Figure for Example 5.4.

Example 5.4: Consider the network in Figure 5.9. We want to use MVA to obtain the mean delay at nodes 1 and 2 with $\mu_1 = \mu_2 = 1$ and $K = 2$.

Solution: We first solve the flow equations for a closed queueing network, which are:

$$\pi_1 = \tfrac{1}{2}\pi_1 + \pi_2 \Rightarrow \tfrac{1}{2}\pi_1 = \pi_2,$$
$$\pi_1 + \pi_2 = 1 \Rightarrow \pi_1 = 2/3, \pi_2 = 1/3,$$
$$\lambda_1 = 2\lambda(2)/3,$$
$$\lambda_2 = \lambda(2)/3.$$

Since $E[N_i(0)] = 0$, then for $K = 1$ we have that:

$$E[T_1(1)] = \frac{1}{\mu_1}\{1 + E[N_1(0)]\} = \frac{1}{\mu_1} = 1,$$

$$E[T_2(1)] = \frac{1}{\mu_2}\{1 + E[N_2(0)]\} = \frac{1}{\mu_2} = 1,$$

$$\lambda(1) = \frac{1}{\pi_1 E[T_1(1)] + \pi_2 E[T_1(1)]} = 1,$$
$$E[N_1(1)] = \lambda(1)\pi_1 E[T_1(1)] = 2/3,$$
$$E[N_2(1)] = \lambda(1)\pi_2 E[T_2(1)] = 1/3.$$

For $K = 2$ we obtain:

$$E[T_1(2)] = \frac{1}{\mu_1}\{1 + E[N_1(1)]\} = \frac{5}{3},$$

$$E[T_2(2)] = \frac{1}{\mu_2}\{1 + E[N_2(1)]\} = \frac{4}{3},$$

$$\lambda(2) = \frac{2}{\pi_1 E[T_1(2)] + \pi_2 E[T_1(2)]} = \frac{9}{7},$$

$$E[N_1(2)] = \lambda(2)\pi_1 E[T_1(2)] = \frac{10}{7},$$

$$E[N_2(2)] = \lambda(2)\pi_2 E[T_2(2)] = \frac{4}{7}.$$

These results are in agreement with those obtained using the convolution algorithm. Also, observe that the throughput at node 1 is $\lambda_1(2) = \pi_1\lambda(2) = 6/7$ and the throughput at node 2 is $\lambda_2(2) = \pi_2\lambda(2) = 3/7$, both of which agree with the results obtained via the convolution algorithm.

For a multiclass network, the arrival theorem becomes the following:

$$E[T_{ir}(K)] = \frac{1}{\mu_{ir}}\{1 + E[N_i(K - e_r)]\} \quad i = 1, \ldots, N; \quad r = 1, \ldots, R.$$

All the steps are then similar to the single-class case.

5.7 NETWORKS WITH POSITIVE AND NEGATIVE CUSTOMERS

A new class of queueing networks called the "generalized networks" (or G-networks) was introduced by Gelenbe et al. (1991). These networks unify different stochastic networks. From a queueing network perspective G-networks have two types of customers: positive customers and negative customers. Positive customers behave like the customers we have discussed in the regular queueing network. Thus, if a positive customer joins a node, it waits until it is served or until it is destroyed by an arriving negative customer. When a negative customer arrives at a node, it destroys a positive customer at the node, if one exists, or it will vanish immediately if there is no customer at the node. Negative customers do not consume the server's time at a node as their actions are taken instantaneously. Thus, a positive customer arrival causes the queue length to increase by one while a negative customer arrival causes the queue length to decrease by one.

An example of a queueing network with negative customers is a computer system with a virus. When there are no viruses, the system behaves as a regular network of queues as previously discussed and analyzed. However, to include the effect of viruses we model the arrival of a virus by the arrival of a negative customer that has the potential to delete or cancel a "positive" customer at the queue upon its arrival. The goal is to find the impact of negative customers on positive customers since it is assumed that negative customers cannot accumulate at a queue and are simply lost if they arrive at an empty queue.

External positive customers arrive at node i according to a Poisson process with rate Λ_i, and negative customers arrive at node i according to a Poisson process with rate λ_i. The time to serve a positive customer at node i is exponentially distributed with mean $1/\mu_i$. After receiving service at node i, a positive customer can do one of three things: move to node j as a positive customer with probability p_{ij}^+, move to node j as a negative customer with probability p_{ij}^-, or leave the network with probability p_{i0}. Thus, for an N-node network, we have that:

$$\sum_{j=1}^{N} p_{ij}^+ + \sum_{j=1}^{N} p_{ij}^- + p_{i0} = 1 \quad 1 \leq i \leq N.$$

It was shown in Gelenbe et al. (1991) that this network has a product-form solution of the form:

$$P[K_1 = k_1, K_2 = k_2, \ldots, K_N = k_N] = \prod_{i=1}^{N} P[K_i = k_i] = \prod_{i=1}^{N} (1 - q_i) q_i^{k_i},$$

where K_i is the number of positive customers at node i and q_i is defined as follows:

$$q_i = \frac{\lambda^+(i)}{\mu_i + \lambda^-(i)},$$

where $\lambda^+(i)$ is the total arrival rate of positive customers at node i and $\lambda^-(i)$ is the total arrival rate of the negative customers at node i. These quantities are given by the following traffic equations:

$$\lambda^+(i) = \Lambda_i + \sum_{j=1}^{N} \mu_j q_j p_{ji}^+,$$

$$\lambda^-(i) = \lambda_i + \sum_{j=1}^{N} \mu_j q_j p_{ji}^-.$$

5.7.1 G-Networks with Signals

In the previous discussion an arriving negative customer causes the length of a nonempty queue to decrease by one. In this subsection we designate a negative customer as a signal that may either destroy one positive customer at the node or move a positive customer to another node instead of destroying it. Let s_{ij}^+ denote the probability that a signal arriving at node i causes a positive customer to move to node j as a positive customer and let s_{ij}^- be the probability that the customer moves to node j as a negative customer. In Chao et al. (1999), it is shown that the solution has a product form stationary distribution as follows:

$$P[K_1 = k_1, K_2 = k_2, \ldots, K_N = k_N] = \prod_{i=1}^{N} P[K_i = k_i]$$

$$= \prod_{i=1}^{N} \left(1 - \frac{\alpha_i^+}{\mu_i + \alpha_i^-}\right) \left(\frac{\alpha_i^+}{\mu_i + \alpha_i^-}\right)^{k_i},$$

provided that:

$$\frac{\alpha_i^+}{\mu_i + \alpha_i^-} < 1 \quad i = 1, \ldots, N,$$

where α_i^+ and α_i^- are customer arrival rate and signal arrival rate at node i, respectively, and are given by the following traffic equations:

$$\alpha_i^+ = \Lambda_i + \sum_{j=1}^{N} \frac{\alpha_j^+ \mu_j}{\mu_j + \alpha_i^-} p_{ji}^+ + \sum_{j=1}^{N} \frac{\alpha_j^+ \alpha_j^-}{\mu_j + \alpha_j^-} s_{ji}^+,$$

$$\alpha_i^- = \lambda_i + \sum_{j=1}^{N} \frac{\alpha_j^+ \mu_j}{\mu_j + \alpha_i^-} p_{ji}^- + \sum_{j=1}^{N} \frac{\alpha_j^+ \alpha_j^-}{\mu_j + \alpha_j^-} s_{ji}^-.$$

Note that Λ_i, λ_i, μ_i, p_{ij}^+, and p_{ij}^- are as previously defined. Also, the departure rate of customers and signals from node are given respectively by:

$$d_i^+ = \frac{\alpha_i^+ \mu_i}{\mu_i + \alpha_i^-},$$

$$d_i^- = \frac{\alpha_i^+ \alpha_i^-}{\mu_i + \alpha_i^-}.$$

5.7.2 Extensions of the G-Network

Other extensions of the G-network have been considered. These include networks with signals and batch service where an arriving signal triggers the instantaneous movement of a customer from node i to node j with probability p_{ij} or with probability r_i, it forces a batch of customers of random size to leave the network, where:

$$r_i = 1 - \sum_j p_{ij}.$$

Other networks are those that can have multiple classes of customers and signals. These examples are discussed in Gelenbe and Pujolle (1998) and in Chao et al. (1999).

5.8   PROBLEMS

5.1   Consider the network shown in Figure 5.10 in which external customers arrive according to a Poisson process with rate γ. There is a single server and the time to serve a customer is exponentially distributed with mean $1/\mu$. After receiving service, a customer leaves the system with probability

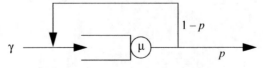

Figure 5.10   Figure for Problem 5.1.

p and rejoins the queue with probability $1 - p$. If we assume that service is on an FCFS basis, find the expected number of customers in the system and the mean time a customer spends in the system.

5.2 Consider the network shown in Figure 5.11, which has three exponential service stations with rates μ_1, μ_2, and μ_3, respectively. External customers arrive at the station labeled Queue 1 according to a Poisson process with rate γ. Let N_1 denote the steady-state number of customers at Queue 1, N_2 the steady-state number of customers at Queue 2, and N_3 the steady-state number of customers at Queue 3. Find the joint PMF $p_{N_1 N_2 N_3}(n_1, n_2, n_3)$.

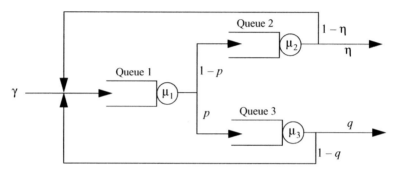

Figure 5.11 Figure for Problem 5.2.

5.3 Consider the acyclic Jackson network of queues shown in Figure 5.12, which has the property that a customer cannot visit a node more than once. Specifically, assume that there are four exponential single-server nodes with service rates μ_1, μ_2, μ_3, and μ_4, respectively, such that external arrivals occur at nodes 1 and 2 according to Poisson processes with rates γ_1 and γ_2, respectively. Upon completion of service at node 1, a customer proceeds to node 2 with probability p, to node 3 with probability r, and to node 4 with probability $1 - p - r$. Similarly, upon completion of service at node 2 a customer proceeds to node 3 with probability q and to node 4 with probability $1 - q$. After receiving service at node 3 or node 4, a customer leaves the system. Find the expected delay at each node.

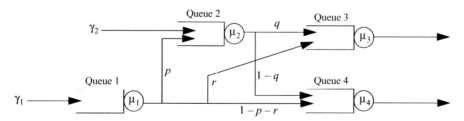

Figure 5.12 Figure for Problem 5.3.

5.4 Consider the closed network of queues shown in Figure 5.13. Assume that the number of customers inside the network is $K = 3$. Find the following:

a. The joint PMF $p_{N_1 N_2 N_3}(n_1, n_2, n_3)$

b. The mean delay at each node using the convolution algorithm

c. The mean delay at each node using the MVA

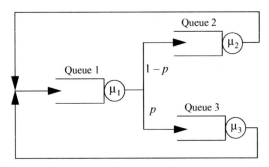

Figure 5.13 Figure for Problem 5.4.

5.5 Consider a closed network with $K = 3$ circulating customers, as shown in Figure 5.14. There is a single exponential server at nodes 1, 2, and 3 with service rates μ_1, μ_2, and μ_3, respectively. After receiving service at node 1 or node 2, a customer proceeds to node 3. Similarly, after receiving service at node 3, a customer goes to node 1 with probability p, and to node 2 with probability $1 - p$. Find the following:

a. The joint PMF $p_{N_1 N_2 N_3}(n_1, n_2, n_3)$

b. The mean delay at each node using the convolution algorithm

c. The mean delay at each node using the MVA

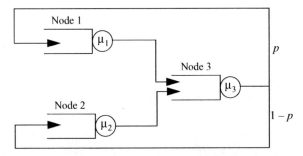

Figure 5.14 Figure for Problem 5.5.

5.6 Consider a tandem queue with two nodes in series as shown in Figure 5.15. Customers arrive at node 1 according to a Poisson process with rate γ, and signals arrive at node 1 according to a Poisson process with rate η. The service time distribution at node i is exponentially distributed with

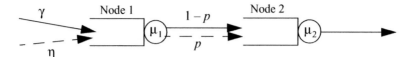

Figure 5.15 Figure for Problem 5.6.

rate μ_i, $i = 1, 2$. When a customer completes service at node 1, it goes to node 2 as a signal with probability p or as a customer with probability $1 - p$. When a signal arrives at node 1, a customer, if at least one is present, is induced to leave the system immediately. A customer that either completes service at node 2 or is induced to move by a signal at node 2 leaves the system immediately. Find the stationary distribution $\pi_{N_1 N_2}(n_1, n_2)$ for the network, where N_1 is the number of customers at node 1 and N_2 is the number of customers at node 2.

5.7 Consider the Problem 5.6 and assume that when a customer arrives at node 2 and is induced by a signal to move, it returns to node 1. Find the stationary distribution $\pi_{N_1 N_2}(n_1, n_2)$ for the network.

5.8 Consider the open network shown in Figure 5.16 with two nodes, labeled node 1 and node 2. External customers arrive at node 1 according to a Poisson process with parameter γ. All external customers arriving at node 1 must return to the node after completing their service at the node. After the second visit to node 1, a customer may depart the system with probability $1 - p$ or go to node 2 with probability p. There is a single exponential server at each node with a service rate of μ_1 at node 1 and μ_2 at node 2.

  a. What is the mean number of visits that a customer makes to node 1 and node 2, respectively?

  b. By defining a suitable state space, obtain a Markov chain for the problem.

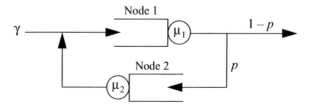

Figure 5.16 Figure for Problem 5.8.

6

APPROXIMATIONS OF QUEUEING SYSTEMS AND NETWORKS

6.1 INTRODUCTION

The exact solutions of many queueing models are available only under restrictive assumptions that are usually not consistent with real life, thereby making approximate solutions a practical necessity. Most of the proposed approximate solutions are based on heavy traffic assumption. That is, the system is operating at such a point that the server is almost continuously busy. Alternatively, the arrival rate of customers is almost equal to the service rate.

Two of the widely used heavy traffic approximate solution techniques are the fluid approximation and the diffusion approximation, both of which are used to provide conversion from discrete state space to continuous state space. The fluid approximation of a queueing system attempts to smooth out the randomness in the arrival and departure processes by replacing these processes with continuous flow functions of time. Thus, fluid approximation can be considered as a first-order approximation that replaces the arrival and departure processes by their mean values, thereby creating a deterministic continuous process.

The diffusion approximation attempts to improve on the fluid approximation by permitting the arrival and departure processes to fluctuate about their means. Thus, it is a second-order approximation that replaces a jump process $N(t)$ with a continuous diffusion process $X(t)$ whose incremental changes $dX(t) = X(t + dt) - X(t)$ are normally distributed with finite variance. The use

Fundamentals of Stochastic Networks, First Edition. Oliver C. Ibe.
© 2011 John Wiley & Sons, Inc. Published 2011 by John Wiley & Sons, Inc.

of diffusion approximation reduces most problems related to system performance to the solution of partial or ordinary differential equation. This connection of diffusion processes to differential equations makes diffusion processes computationally and analytically attractive. In general, using a differential equation to represent a process has the advantage that a differential equation is convenient for the mathematical treatment of boundary conditions and initial value problems because the boundary conditions appear as subsidiary conditions that do not explicitly appear in the differential equation. Another way to view the diffusion approximation is that proposed by Gaver (1971) as a backlog model. A backlog exists when the number of customers in the system is large, which is consistent with our assumption that the queue length is much larger than one.

The justification of the diffusion approximation can be explained as follows. Heavy traffic approximations in queueing system analysis are based on the assumption that the number of customers in the system is much larger than one most of the time. In queueing systems, the number of customers in queue is a stochastic process with discontinuous jumps. Thus, when the number of customers in the system is large, it is reasonable to replace these discontinuities by smooth continuous functions of time.

Harrison (1985) observed that under heavy traffic conditions the scheduling problem in multiclass queueing networks can be approximated by a control problem involving the Brownian motion. Thus, the Brownian approximation is commonly used in manufacturing systems and other queueing systems that deal with multiple classes of customers with different priorities.

In this chapter we discuss how single-server queueing systems and queueing networks are approximated by the fluid flow process and the diffusion process. As stated earlier, the fundamental assumption is that the system or network is operating close to system capacity so that the number of customers in the system is much greater than one. We also discuss Brownian motion approximation of the G/G/1 queue.

6.2 FLUID APPROXIMATION

Fluid approximation of a queueing system involves treating the system as a continuous fluid flow rather than a discrete customer flow. Thus, the discrete jump processes associated with customer arrivals and departures are replaced by smooth continuous functions of time. The approximation is based on the assumption that there is at least one customer in the system most of the time. This means that it is a heavy traffic approximation in which the magnitude of the original discontinuities is small relative to the average value of the function.

6.2.1 Fluid Approximation of a G/G/1 Queue

Let $A(t)$ denote the cumulative number of arrivals up to time t, and let $D(t)$ denote the cumulative number of departures up to time t. Then

$Q(t) = A(t) - D(t)$ is the number of customers in the system at time t. Let $E[A(t)]$ denote the expected value of $A(t)$, and let $E[D(t)]$ denote the expected value of $D(t)$. When $A(t)$ is large compared to unity, then we expect only a small percentage of deviations from its average values. That is, according to the strong law of large numbers,

$$P\left[\lim_{t\to\infty}\frac{A(t)-E[A(t)]}{E[A(t)]}=0\right]=1.$$

This means that a heavy traffic approximation of the queueing system is to replace the arrival process by its average value as a function of time. Similarly, we can replace the departure process $D(t)$ by its mean value $E[D(t)]$ and the number of customers in the system can be approximated by:

$$Q(t) = E[A(t)] - E[D(t)].$$

Assume that customers arrive at an average rate of λ customers per unit time (or the mean interarrival time is $1/\lambda$ time units), and customers are served at an average rate of μ customers per unit time (or the mean time to serve a customer is $1/\mu$ time units). Then we can approximate $A(t)$ by λt and we can approximate $D(t)$ by μt. If we assume that the number of customers in the system at time $t = 0$ in $Q(0)$, then we have that:

$$Q(t) = Q(0) + (\lambda - \mu)t.$$

If $\lambda < \mu$, $Q(t)$ eventually becomes zero, but if $\lambda > \mu$, the number of customers in the system increases with time. In the event that the arrival rate is a time-dependent function $\lambda(t)$ and the departure rate is a time-dependent function $\mu(t)$, then we have that:

$$Q(t) = Q(0) + \int_0^t \{\lambda(u) - \mu(u)\} du.$$

A rush hour traffic model was introduced by Newell (1971) to model the time-dependent arrival rate as follows:

$$\lambda(t) = \lambda(t_1) - \alpha(t - t_1)^2,$$

where t_1 is the point at which the function achieves it maximum and α is some constant, as shown in Figure 6.1.

Thus, we have that:

$$\alpha = -\frac{1}{2}\frac{d^2\lambda(t)}{dt^2}\bigg|_{t=t_1}.$$

Let $t_0 < t_1 < t_2 < t_3$ be such that $\lambda(t_0) = \lambda(t_2) = \mu$, which is the service rate, and $Q(t_3) = 0$. Thus, we have that:

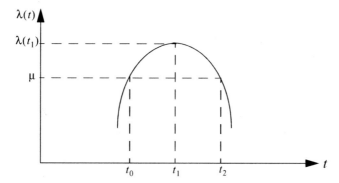

Figure 6.1   A rush hour model.

$$\mu = \lambda(t_1) - \alpha(t_0 - t_1)^2 = \lambda(t_1) - \alpha(t_2 - t_1)^2,$$

from which we obtain:

$$t_0 = t_1 - \left\{ \frac{\lambda(t_1) - \mu}{\alpha} \right\}^{1/2},$$

$$t_2 = t_1 + \left\{ \frac{\lambda(t_1) - \mu}{\alpha} \right\}^{1/2},$$

$$t_2 - t_0 = 2 \left\{ \frac{\lambda(t_1) - \mu}{\alpha} \right\}^{1/2}.$$

Thus, we have that:

$$\lambda(t) - \mu = \lambda(t_1) - \alpha(t - t_1)^2 - \left\{ \lambda(t_1) - \alpha(t_0 - t_1)^2 \right\} = \alpha \left\{ (t_0 - t_1)^2 - (t - t_1)^2 \right\},$$

and:

$$\lambda(t) - \mu = \lambda(t_1) - \alpha(t - t_1)^2 - \left\{ \lambda(t_1) - \alpha(t_2 - t_1)^2 \right\} = \alpha \left\{ (t_2 - t_1)^2 - (t - t_1)^2 \right\}.$$

This suggests that we can write:

$$\lambda(t) - \mu = K(t - t_0)(t - t_2).$$

From this and an earlier result we have that:

$$\left. \frac{d^2 \lambda(t)}{dt^2} \right|_{t=t_1} = -2\alpha = 2K \Rightarrow K = -\alpha,$$

which gives:

$$\lambda(t) - \mu = \alpha(t - t_0)(t_2 - t).$$

Thus,

$$Q(t) = Q(0) + \int_0^t \{\lambda(u) - \mu(u)\} du = Q(0) + \int_0^t \{\lambda(u) - \mu\} du$$

$$= Q(0) + \int_0^t \{\alpha(u - t_0)(t_2 - u)\} du$$

$$= Q(0) + \alpha(t - t_0)^2 \left\{\frac{t_2 - t_0}{2} - \frac{t - t_0}{3}\right\}.$$

Assume that $Q(0) = 0$, then since $Q(t_3) = 0$, we have that:

$$t_2 - t_0 = \frac{2}{3}(t_3 - t_0).$$

Thus, we have that:

$$Q(t) = \alpha(t - t_0)^2 \left\{\frac{t_2 - t_0}{2} - \frac{t - t_0}{3}\right\} = \frac{\alpha}{3}(t - t_0)^2 \{(t_3 - t_0) - (t - t_0)\}$$

$$= \frac{\alpha}{3}(t - t_0)^2 (t_3 - t).$$

Finally, under the assumption that $Q(0) = 0$, the total delay in the system between t_0 and t_3 is given by:

$$T = \int_{t_0}^{t_3} Q(u) du = \frac{\alpha}{3} \int_{t_0}^{t_3} (u - t_0)^2 (t_3 - u) du.$$

Let $v = u - t_0$, which implies that $dv = du$ and:

$$T = \frac{\alpha}{3} \int_0^{t_3 - t_0} v^2(t_3 - t_0 - v) dv = \frac{\alpha}{3} \int_0^{t_3 - t_0} \{(t_3 - t_0)v^2 - v^3\} dv = \frac{\alpha}{36}(t_3 - t_0)^4.$$

Recall that:

$$t_2 - t_0 = 2\left\{\frac{\lambda(t_1) - \mu}{\alpha}\right\}^{1/2} = \frac{2}{3}(t_3 - t_0) \Rightarrow t_3 - t_0 = 3\left\{\frac{\lambda(t_1) - \mu}{\alpha}\right\}^{1/2},$$

which gives the total delay as follows:

$$T = \frac{9[\lambda(t_1) - \mu]^2}{4\alpha}.$$

Note that the preceding results are based on the deterministic assumptions that we made about the system and are dependent on the state of the system at specific points in time. This means that there are limitations on the usefulness of these results.

6.2.2   Fluid Approximation of a Queueing Network

A fluid network is a deterministic network model in which continuous flows
are processed and circulated among a set of nodes. As in the single-server case,
the notion of discrete customers is replaced by the notion of fluids or customer
mass. We consider only the single-customer class network.

Consider a network with K single-server stations (or queues), labeled
$1, 2, \ldots, K$, each of which has an infinite capacity. Let $\lambda_{0k}(t)$ denote the cumulative external arrivals (or inflow) to station k during the time interval $[0, t]$,
and let $\mu_k(t)$ be the cumulative customer flow out of station k during the interval $[0, t]$, $k = 1, 2, \ldots, K$. After passing through station k, a fraction p_{kj} of the
flow goes directly to station j and a fraction $1 - \sum_j p_{kj}$ leaves the network. Let
$P = [p_{kj}]$ be a $K \times K$ routing matrix.

Let $Q_i(t)$ be the fluid level at station i at time t. Then we have that:

$$Q_i(t) = Q_i(0) + \lambda_{0i}(t) + \sum_{j=1}^{K} \mu_j(t) p_{ji} - \mu_i(t),$$

where $Q_i(0)$ is the initial fluid level at station i. Let $Q(t)$ be the K-dimensional
vector-valued function $Q(t) = \{Q_1(t), Q_2(t), \ldots, Q_K(t)\}$, let $\lambda_0(t) = \{\lambda_{01}(t),$
$\lambda_{02}(t), \ldots, \lambda_{0K}(t)\}$ be the K-dimensional vector external arrival function, and
let $\mu(t) = \{\mu_1(t), \mu_2(t), \ldots, \mu_K(t)\}$ be the K-dimensional vector outflow function.
Then the preceding equation can be expressed in the form:

$$Q(t) = Q(0) + \lambda_0(t) + P^T \mu(t) - \mu(t) = Q(0) + \lambda_0(t) - \left[I - P^T\right] \mu(t),$$

where P^T denotes the transpose of P and I is the identity matrix. Note that
$P = [p_{kj}]$ is a substochastic matrix; that is, each row may not sum to one since
customers (or flows) are allowed to leave the network from any station.
However, if the network is closed, then P is a stochastic matrix. As in the
single-queue case, the model developed here is a purely deterministic one as
there is no random phenomenon in the network.

6.3   DIFFUSION APPROXIMATIONS

Fluid approximations are deterministic, continuous approximations to stochastic discrete networks that replace discrete jobs moving stochastically
through the network with a continuous deterministic flow. The diffusion
approximation improves the fluid approximation by introducing variations
about the mean in the arrival and departure processes that are normally
distributed.

6.3.1 Diffusion Approximation of a G/G/1 Queue

A diffusion process is a continuous-state, continuous-time Markov process whose state changes continually, but only small changes occur in small time intervals. The mean and variance of the state change during a small interval of time are finite. Consider a single-server queue where the arrival times of customers are $0 < \tau_1 < \tau_2 < \dots$ and the departure times of these customers are $0 < \tau'_1 < \tau'_2 < \dots$. Let $A(t)$ denote the cumulative number of arrivals up to time t, and let $D(t)$ denote the cumulative number of departures up to time t. Then $Q(t) = A(t) - D(t)$ is the number of customers in the system at time t. Let $1/\lambda$ and σ_a^2 be the mean and variance of the interarrival times, respectively; and let $1/\mu$ and σ_s^2 be the mean and variance of the service times, respectively. In Heyman (1975) it is shown that the mean and variance of $A(t)$ and $D(t)$ can be approximated as follows:

$$E[A(t)] \approx \lambda t, \quad \sigma_{A(t)}^2 \approx \lambda^3 \sigma_a^2 t,$$
$$E[D(t)] \approx \mu t, \quad \sigma_{D(t)}^2 \approx \mu^3 \sigma_s^2 t.$$

The change in $Q(t)$ between time t and $t + \Delta t$ is given by

$$Q(t+\Delta t) - Q(t) = [A(t+\Delta t) - A(t)] - [D(t+\Delta t) - D(t)] \equiv Q(\Delta t).$$

The incremental time $X_i = \tau_i - \tau_{i-1}$ is the interval between the arrival of the $(i-1)$th customer and the arrival of the ith customer. We assume that the interarrival times X_i are independent and identically distributed with mean $E[X] = 1/\lambda$ and variance σ_X^2. If we assume that the queue is never empty so that there is no idle time between the completion of the service of the $(i-1)$th customer and the start of the service of the ith customer, then the service times are given by:

$$S_i = \tau'_i - \tau'_{i-1},$$

and it is assumed that the service times are independent and identically distributed with mean $E[S] = 1/\mu$ and variance σ_s^2.

We define $A(\Delta t) = A(t + \Delta t) - A(t)$ and $D(\Delta t) = D(t + \Delta t) - D(t)$. From the central limit theorem, if Δt is sufficiently large so that many events (arrivals and departures) take place between t and $t + \Delta t$ and $Q(t)$ does not become zero in this interval, then $Q(\Delta t) = A(\Delta t) - D(\Delta t)$ can be approximately normally distributed with mean and variance given by:

$$E[Q(\Delta t)] = (\lambda - \mu)\Delta t = \beta \Delta t,$$
$$\sigma_{Q(\Delta t)}^2 = \sigma_{A(\Delta t)}^2 + \sigma_{D(\Delta t)}^2 = \left(\lambda^3 \sigma_a^2 + \mu^3 \sigma_s^2\right)\Delta t = \alpha \Delta t.$$

In general, α and β depend on the initial value of $N(t)$ and so are given by:

$$\beta(q_0) = \lim_{\Delta t \to 0} \frac{E[Q(t+\Delta t) - Q(t)|Q(t) = q_0]}{\Delta t},$$

$$\alpha(q_0) = \lim_{\Delta t \to 0} \frac{\text{Var}[Q(t+\Delta t) - Q(t)|Q(t) = q_0]}{\Delta t}.$$

Thus, the continuous process $X(t)$ that is used to approximate the discrete-valued process $Q(t)$ is defined by the following stochastic equation:

$$dX(t) = \beta dt + B(t)\sqrt{\alpha}dt,$$
$$X(t) \geq 0,$$

where $B(t)$ is a white Gaussian process with zero mean and unit variance. We know that given the initial value of x_0, the conditional probability density function (PDF) at time t is given by:

$$f_{X(t)|X(0)}(x, t|x_0) = \frac{1}{\sqrt{2\pi\alpha t}} \exp\left\{-\frac{(x - x_0 - \beta t)^2}{2\alpha t}\right\}.$$

Thus, we have that:

$$\frac{\partial f_{X(t)|X(0)}(x, t|x_0)}{\partial t} = -\frac{f_{X(t)|X(0)}(x, t|x_0)}{2t}\left[1 - \frac{(x - x_0 - \beta t)^2}{\alpha t} - \frac{2\beta(x - x_0 - \beta t)}{\alpha}\right],$$

$$\frac{\partial f_{X(t)|X(0)}(x, t|x_0)}{\partial x} = -\frac{(x - x_0 - \beta t)}{\alpha t} f_{X(t)|X(0)}(x, t|x_0),$$

$$\frac{\partial^2 f_{X(t)|X(0)}(x, t|x_0)}{\partial x^2} = -\frac{f_{X(t)|X(0)}(x, t|x_0)}{\alpha t}\left[1 - \frac{(x - x_0 - \beta t)^2}{\alpha t}\right].$$

From these partial derivatives we see that:

$$\frac{\partial f_{X(t)|X(0)}(x, t|x_0)}{\partial t} = \frac{\alpha}{2} \frac{\partial^2 f_{X(t)|X(0)}(x, t|x_0)}{\partial x^2} - \beta \frac{\partial f_{X(t)|X(0)}(x, t|x_0)}{\partial x},$$

which is the classic *diffusion equation*. The steady-state distribution of $X(t)$ is obtained by setting $\partial f/\partial t = 0$, replacing $f_{X(t)|X(0)}(x, t|x_0)$ by $f_X(x)$ along with the requirement:

$$\int_0^\infty f_X(x)dx = 1.$$

That is,

$$0 = \frac{\alpha}{2} \frac{d^2 f_X(x)}{dx^2} - \beta \frac{df_X(x)}{dx},$$
$$x > 0.$$

This gives to the result:

$$f_X(x) = \begin{cases} -\gamma \exp(\gamma x) & x \geq 0 \\ 0 & x < 0, \end{cases}$$

where $\gamma = 2\beta/\alpha = 2(\lambda - \mu)/(\lambda + \mu) = -2(1 - \rho)/(1 + \rho)$, $\rho = \lambda/\mu < 1$. Thus, the limiting distribution of $\{X(t)|X(0) = x_0\}$ is an exponential distribution with mean:

$$\lim_{t \to \infty} E[X(t)|X(0) = x_0] = -\frac{1}{\gamma} = \frac{\alpha}{-2\beta} = \frac{\lambda^3 \sigma_a^2 + \mu^3 \sigma_s^2}{2(\mu - \lambda)}.$$

Having obtained the equilibrium state in terms of the PDF $f_X(x)$, we would like to obtain an approximate expression for the distribution of the number of customers in the system. Let the probability mass function (PMF) of the number of customers in the system be $p_Q(n)$ whose approximate value is given by:

$$\hat{p}_Q(n) = \int_1^{n+1} f_X(x)dx = \{1 - \eta\}\eta^n \quad n = 0, 1, 2, \ldots,$$

where $\eta = \exp(\gamma)$. Thus, the expression is similar to the mean number of customers in an M/M/1 queue.

Let $Y(t)$ denote the total workload at time t; that is, $Y(t)$ is the total amount of work brought to the server in $(0, t]$. Assume that $Y(0) = 0$. Then:

$$Y(t) = S_1 + S_2 + \ldots + S_{A(t)} \quad t > 0.$$

Because the service times S_i are independent and identically distributed, we have that:

$$E[Y(t)] = E[S]E[A(t)] \approx \rho t,$$
$$\sigma_{Y(t)}^2 = E[A(t)]\sigma_s^2 + (E[S])^2 \sigma_a^2 = \lambda\{\sigma_s^2 + \rho^2 \sigma_a^2\}t.$$

Let $W(t)$ denote the virtual waiting time, which is the waiting time in queue of a customer arriving at time t. The relationship between $Y(t)$ and $W(t)$ is given by:

$$W(t + \Delta t) - W(t) = Y(t + \Delta t) - Y(t) - \Delta t,$$

where $W(t) \geq \Delta t$. Thus, from the preceding results we have that:

$$\lim_{\Delta t \to \infty} \left\{ \frac{E[W(t + \Delta t) - W(t)]}{\Delta t} \right\} = \lim_{\Delta t \to \infty} \left\{ \frac{E[Y(t + \Delta t) - Y(t)]}{\Delta t} \right\} - 1 = \rho - 1,$$

$$\lim_{\Delta t \to \infty} \left\{ \frac{\text{Var}[W(t + \Delta t) - W(t)]}{\Delta t} \right\} = \lim_{\Delta t \to \infty} \left\{ \frac{\text{Var}[Y(t + \Delta t) - Y(t) - \Delta t]}{\Delta t} \right\}$$

$$= \lambda\{\sigma_s^2 + \rho^2 \sigma_a^2\}.$$

The virtual waiting time process $\{W(t), t \geq 0\}$ can be approximated by a Brownian motion with infinitesimal mean and variance given by (Gaver 1968):

$$\beta = \rho - 1,$$
$$\alpha = \lambda \{\sigma_s^2 + \rho^2 \sigma_a^2\}.$$

Thus, the limiting distribution of the virtual waiting time is exponentially distributed with mean given by:

$$\lim_{t \to \infty} E[W(t)] = \frac{\alpha}{-2\beta} = \frac{\lambda \{\sigma_s^2 + \rho^2 \sigma_a^2\}}{2(1-\rho)}.$$

The values of α and β depend on the type of queueing model used. We consider two examples.

Example 6.1: M/M/1 Queue

In this case, $\sigma_a^2 = 1/\lambda^2$ and $\sigma_s^2 = 1/\mu^2$. Thus, we have that:

$$\lim_{t \to \infty} E[W(t)] = \frac{\lambda \{\sigma_s^2 + \rho^2 \sigma_a^2\}}{2(1-\rho)} = \frac{\lambda \{(1/\mu)^2 + (1/\mu)^2\}}{2(1-\rho)} = \frac{\rho}{\mu(1-\rho)},$$

which is the result we obtained in Chapter 3 for the M/M/1 queue.

Example 6.2: M/G/1 Queue

In this case, $\sigma_a^2 = 1/\lambda^2$. Thus, we have that:

$$\lim_{t \to \infty} E[W(t)] = \frac{\lambda \{\sigma_s^2 + \rho^2 \sigma_a^2\}}{2(1-\rho)} = \frac{\lambda \{\sigma_s^2 + (E[S])^2\}}{2(1-\rho)} = \frac{\lambda E[S^2]}{2(1-\rho)} = \frac{\rho \{1 + C_s^2\} E[S]}{2(1-\rho)},$$

which is the Pollaczek–Khinchin formula for the mean waiting time in an M/G/1 queue that we obtained in Chapter 3.

6.3.2 Brownian Approximation for a G/G/1 Queue

The Brownian motion is an example of a diffusion process. A Brownian motion $\{W(t), t \geq 0\}$ is a stochastic process that models random continuous motion. It is considered to be the continuous-time analog of the random walk and can also be considered as a continuous-time Gaussian process with independent increments. In particular, the Brownian motion has the following properties:

1. $W(0) = 0$; that is, it starts at zero.
2. $W(t)$ is continuous in $t \geq 0$; that is, it has continuous sample paths with no jumps.

3. It has both stationary and independent increments.
4. For $0 \le s < t$, the random variable $W = W(t) - W(s)$ has a normal distribution with mean 0 and variance $\sigma_W^2 = \sigma^2(t-s)$. That is, $W \sim N(0, \sigma^2(t-s))$.

The Brownian motion is an important building block for modeling continuous-time stochastic processes. In particular, it has become an important framework for modeling financial markets. The path of a Brownian motion is always continuous, but it is nowhere smooth; consequently, it is nowhere differentiable. The fact that the path is continuous means that a particle in Brownian motion cannot jump instantaneously from one point to another.

Because $W(0) = 0$, then according to property 3,

$$W(t) = W(t) - W(0) \sim N\left(0, \sigma^2(t-0)\right) = N\left(0, \sigma^2 t\right).$$

Thus, $W(t - s) \sim N(0, \sigma^2(t - s))$; that is, $W(t) - W(s)$ has the same distribution as $W(t - s)$. This also means that another way to define a Brownian motion $\{W(t), t \ge 0\}$ is that it is a process that satisfies conditions 1, 2, and 3 along with the condition $W(t) \sim N(0, \sigma^2 t)$.

There are many reasons for studying the Brownian motion. As stated earlier, it is an important building block for modeling continuous-time stochastic processes because many classes of stochastic processes contain the Brownian motion. It is a Markov process, a Gaussian process, a martingale, a diffusion process, as well as a Levy process. Over the years it has become a rich mathematical object. For example, it is the central theme of stochastic calculus.

A Brownian motion is sometimes called a *Wiener process*. A sample function of the Brownian motion is shown in Figure 6.2.

Let $B(t) = W(t)/\sigma$. Then $E[B(t)] = 0$ and $\sigma_{B(t)}^2 = 1$. The stochastic process $\{B(t), t \ge 0\}$ is called the *standard Brownian motion*, which has the property that when sampled at regular intervals it produces a symmetric random walk. Note that $B(t) = N(0, t)$. In the remainder of this chapter we refer to the Weiner process $\{W(t), t \ge 0\}$ as the *classical Brownian motion* and use the two terms interchangeably.

Another property of the Brownian motion is *scale invariance*. A stochastic process $\{X(t), t \in R\}$ is defined to be scale invariant with scaling exponent H and scale λ if:

$$X(\lambda t) \cong \lambda^H X(t),$$

$W(t)$

Figure 6.2   Sample function of the Brownian motion process.

where $\cong$ means identical distribution. Brownian motion is scale invariant, provided we rescale time accordingly. This means that the statistical properties of a Brownian motion are the same at all time scales. Thus, the process:

$$Y(t) = kB(t/k^2) \quad t \geq 0,$$

is another Brownian motion that is indistinguishable from the Brownian motion $B(t)$ except for the scale. Similarly, we have that:

$$B(\lambda t) \cong \lambda^2 B(t).$$

6.3.2.1 *Brownian Motion with Drift* Brownian motion is used to model stock prices. Because stock prices do not generally have a zero mean, it is customary to include a *drift* measure that makes the following model with a *drift rate* $\mu > 0$ a better model than the classical Brownian motion:

$$Y(t) = \mu t + W(t) \quad t \geq 0,$$

where $W(t)$ is the classical Brownian motion. $Y(t)$ is called a Brownian motion with drift. Note that $E[Y(t)] = \mu t$ and $\sigma^2_{Y(t)} = \sigma^2 t$, which means that $Y(t) \sim N(\mu t, \sigma^2 t)$. Note also that we can express $Y(t)$ in terms of the standard Brownian motion as follows:

$$Y(t) = \mu t + \sigma B(t) \quad t \geq 0$$

6.3.2.2 *Reflected Brownian Motion* Let $\{B(t), t \geq 0\}$ be a standard Brownian process. The stochastic process:

$$R(t) = |B(t)| = \begin{cases} B(t) & \text{if } B(t) \geq 0 \\ -B(t) & \text{if } B(t) < 0, \end{cases}$$

is called a reflected Brownian motion. The process moves back to the positive values whenever it reaches the zero value. Sometimes the process is defined as follows:

$$R(t) = B(t) - \inf_{0 \leq s \leq t} B(s) \quad t \geq 0.$$

Figure 6.3 shows the sample path of the reflected Brownian motion. The dotted line segments below the time axis are parts of the original Brownian motion;

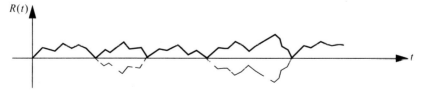

Figure 6.3 Sample function of the reflected Brownian motion process.

these have now been reflected on the time axis to produce the reflected Brownian motion.

Because of the symmetric nature of the PDF of $B(t)$, the cumulative distribution function (CDF) and PDF of $R(t)$ are given by:

$$
\begin{aligned}
F_{R(t)}(r) &= P[R(t) \leq r] = P[\|B(t)\| \leq r] = P[-r \leq B(t) \leq r] \\
&= P[B(t) \leq r] - P[-r \leq B(t)] = P[B(t) \leq r] - \{1 - P[B(t) \leq r]\} \\
&= 2P[B(t) \leq r] - 1,
\end{aligned}
$$

$$
\begin{aligned}
f_{R(t)}(r) &= \frac{dF_{R(t)}(t)}{dr} = 2 f_{B(t)}(r) \\
&= \frac{2}{\sqrt{2\pi t}} \exp\left\{ -\frac{r}{2t} \right\}^2, \quad r \geq 0.
\end{aligned}
$$

Since $B(t) \sim N(0, t)$, the expected value of $R(t)$ is given by:

$$
E[R(t)] = \int_0^\infty r f_{R(t)}(t) dr = 2 \int_0^\infty \frac{r}{\sqrt{2\pi t}} \exp\left\{ -\frac{r^2}{2t} \right\} dr = 2 \int_0^\infty \frac{r}{\sqrt{2\pi t}} \exp\left\{ -\left(\frac{r}{\sqrt{2t}} \right)^2 \right\} dr.
$$

Let $y = r/\sqrt{2t} \Rightarrow dr = dy\sqrt{2t}$, and we obtain:

$$
E[R(t)] = \frac{2\sqrt{2t}}{\sqrt{\pi}} \int_0^\infty y \exp(-y^2) dy = \frac{2\sqrt{2t}}{2\sqrt{\pi}} \int_0^\infty \exp(-u) dr = \sqrt{\frac{2t}{\pi}}.
$$

Similarly, the variance of $R(t)$ is given by:

$$
\sigma_{R(t)}^2 = E\left[\{R(t)\}^2 \right] - \{E[R(t)]\}^2 = E\left[\{B(t)\}^2 \right] - \frac{2t}{\pi} = t - \frac{2t}{\pi} = \left\{ 1 - \frac{2}{\pi} \right\} t.
$$

6.3.2.3 Scaled Brownian Motion

Let $B(t)$ be the standard Brownian motion. The process:

$$
X(t) = \sqrt{\alpha} B(t/\alpha) \quad t \geq 0,
$$

where α is an arbitrary positive constant, is called a scaled Brownian motion. The time scale of the Brownian motion is reduced by a factor α and the magnitude of the process is multiplied by a factor $\sqrt{\alpha}$. Thus, the speed of the process is α times as fast as the Brownian motion while the magnitude is amplified $\sqrt{\alpha}$ times. The expected value and variance of $X(t)$ are given by:

$$
E[X(t)] = E\left[\sqrt{\alpha} B(t/\alpha) \right] = \sqrt{\alpha} E[B(t/\alpha)] = 0,
$$

$$
\sigma_{X(t)}^2 = \text{Var}\left(\sqrt{\alpha} B(t/\alpha) \right) = \left(\sqrt{\alpha} \right)^2 \text{Var}(B(t/\alpha)) = \alpha(t/\alpha) = t.
$$

Thus, the scaled Brownian motion has the same mean and variance as the Brownian motion. This is consistent with the scale invariance property of the Brownian motion.

6.3.2.4 *Functional Central Limit Theorem* In Chapter 1 it was stated that the central limit theorem provides an approximation to the behavior of sums of random variables. It uses the first two moments of the random variables, namely, the mean and the variance. A similar approximation called the functional central limit theorem (FCLT) is used for random processes using the mean and the autocovariance functions. FCLT is also called the *Donsker's theorem*.

The Brownian motion can be thought of as a continuous-time approximation of a random walk where the step size is scaled to become smaller in order for the approximation to have continuous paths, and the rate at which the steps are taken is speeded up. Consider the discrete-time independent and identically distributed sequence $\{X(k); k = 1, 2, \ldots\}$, and let S_n be defined as follows:

$$S_0 = 0,$$

$$S_n = \sum_{i=1}^{n} X(i).$$

Thus, S_n is a random walk. Assume that $E[X(k)] = \mu_X$ and $\mathrm{Var}(X(k)) = \sigma_x^2$, where $0 < \sigma_X^2 < \infty$. This implies that $E[S_n] = n\mu_X$ and $\mathrm{Var}(S_n) = \sigma_x^2$. Let $Y_n(t)$ be a continuous-time process defined as follows:

$$Y_n(t) = \frac{S_{[nt]}}{\sqrt{n}} \quad t \geq 0,$$

where $[nt]$ is the integer part of nt. The FCLT states that $Y_n(t)$ converges in distribution to the standard Brownian motion process; that is,

$$\lim_{n \to \infty} Y_n(t) \to B(t).$$

The theorem states that $Y_n(t)$, which is the continuous-time representation of the random walk S_n, that is obtained by rescaling (or shrinking) the location to the value $n^{-1/2}S_n$ and rescaling the time scale such that the walk takes $[nt]$ steps in time t, converges in distribution to the standard Brownian motion as n becomes large and the steps become small and more frequent.

6.3.2.5 *Brownian Motion Approximation of the G/G/1 Queue* A queueing system is defined to be stable if the input rate is less than the output rate, which means that the traffic intensity $\rho = \lambda/\mu$ is strictly less than one, where λ is the arrival rate of customers and μ is the service rate of customers. (Alternatively, $1/\lambda$ is the mean interarrival time and $1/\mu$ is the mean service time.)

Consider a G/G/1 queue. Let $A(k)$ denote the number of arrivals at time k and let $D(k)$ denote the number of departures at time k. Then the number of customers in the system at time k, $Q(k)$, evolves according to the equation:

$$Q(k+1) = \max\{Q(k) + A(k) - D(k), 0\}.$$

Assume that $Q(0) = 0$ and that we are interested in the steady-state probability that the number of customers in the system exceeds a certain value M. Solving the preceding equation iteratively yields:

$$
\begin{aligned}
Q(1) &= \max\{A(0) - D(0), 0\}, \\
Q(2) &= \max\{Q(1) + A(1) - D(1), 0\} \\
&= \max\{A(1) + A(0) - D(1) - D(0), A(1) - D(1), 0\}, \\
Q(3) &= \max\{Q(2) + A(2) - D(2), 0\} \\
&= \max\{A(2) + D(1) + A(0) - [D(2) + D(1) + D(0)], \\
&\qquad A(2) + A(1) - [D(2) + D(1)], A(2) - D(2), 0\}.
\end{aligned}
$$

We assume that $A(k)$ and $D(k)$ are stationary processes and we define,

$$V(k) = \sum_{j=0}^{k-1}\{A(j) - D(j)\},$$

$$E[V(k)] = k\{E[A(k) - Dk]\} = k\{\lambda - \mu\},$$

where we assume that $E[A(k)] = \lambda$ and $E[D(k)] = \mu$, we have that:

$$
\begin{aligned}
Q(3) &= \max\{A(2) + A(1) + A(0) - [D(2) + D(1) + D(0)], A(2) + A(1) \\
&\qquad - [D(2) + D(1), A(2) - D(2), 0\} \\
&= \max\{V(3), V(3) - V(1), V(3) - V(2), 0\} \\
&= \max\{V(3), V(3) - V(1), V(3) - V(2), V(3) - V(3)\} \\
&= V(3) - \min_{1 \le i \le 3} V(i).
\end{aligned}
$$

In a similar manner it can be shown that:

$$Q(k) = V(k) - \min_{1 \le i \le k} V(i).$$

We consider the system in heavy traffic during which λ approaches μ, but $\lambda < \mu$. Let $V_n(t)$ be the continuous-time process that is defined by:

$$V_n(t) = \frac{V(\lfloor nt \rfloor)}{\sqrt{n}},$$

where $\lfloor x \rfloor$ is the integer part of x. Using the FCLT arguments it can be shown that $V_n(t)$ converges to a Brownian motion process $W(t)$ with drift η and infinitesimal variance σ^2, where η and σ^2 are given by:

$$\eta = \lim_{n \to \infty} \left\{ (\mu - \lambda)\sqrt{n} \right\},$$

$$\sigma^2 = \lim_{k \to \infty} \left\{ \frac{\mathrm{Var}(V(k))}{k} \right\} < \infty.$$

That is,

$$\lim_{t \to \infty} \frac{V(\lfloor nt \rfloor)}{\sqrt{n}} = \lim_{t \to \infty} V_n(t) \Rightarrow W(t).$$

If we assume that:

$$X(t) = \lim_{n \to \infty} \frac{Q_n(t)}{\sqrt{n}}$$

exists, then we have that:

$$X(t) = W(t) - \inf_{0 \leq u \leq f} W(u),$$

which is a reflected Brownian motion with drift, where $W(t)$ is the classical Brownian motion. As we discussed earlier, the PDF of $X(t)$ is given by:

$$f_{X(t)}(x) = 2 f_{W(t)}(x) = \frac{2}{\sigma\sqrt{2\pi t}} \exp\left\{ -\frac{(x - \eta t)^2}{2\sigma^2 t} \right\}, \quad x \geq 0.$$

6.3.3 Diffusion Approximation of Open Queueing Networks

Consider a single-class network with N nodes labeled $1, 2, \ldots, N$. Each node has a single server that serves customers in the order in which they arrive. The service times at node n are independent and identically distributed with mean $1/\mu_n$ and variance σ_n^2, $n = 1, \ldots, N$. Let node 0 denote the point of entry of customers into the network and let node $N + 1$ denote the node through which customers leave the network. Also, let $p_{n,m}$ denote the probability that a customer that finishes receiving service at node n will next proceed to node m for more service. We assume that $p_{n,n} = 0$; that is, no immediate feedback of customers is allowed. This is an example of a *generalized Jackson network*. Unlike the classical Jackson network where the interarrival times and service times are exponentially distributed, the interarrival times and the service times of the generalized Jackson network have a general distribution. Note that $p_{0,n}$ is the probability that an arriving customer enters the network through node n, and $p_{n,N+1}$ is the probability that a customer leaves the network after receiving service at node n. Thus, we have that:

$$\sum_{m=1}^{N} p_{0,m} = 1,$$

$$\sum_{m=1}^{N+1} p_{n,m} = 1 \quad 1 \leq n \leq N,$$

$$p_{n,m} \geq 0, \quad 0 \leq n \leq N, 1 \leq m \leq N+1.$$

Let $A_n(t)$ denote the cumulative number of arrivals at node n up to time t, and let $D_n(t)$ denote the cumulative number of departures from node n up to time t. Assuming that immediate feedback to a node is not allowed, we have that:

$$A_n(t) = D_n(t) p_{0,n} + \sum_{k=1,k \neq n}^{N} D_k(t) p_{k,n} \quad 1 \leq n \leq N.$$

Note that although we have assumed that all external arrivals enter the network through node 0, we can permit external arrivals to enter the network through any node n according to a renewal process with rate $\lambda_n, n = 1, \ldots, N$. In this case the preceding equation becomes:

$$A_n(t) = \lambda_n t + \sum_{k=1,k \neq n}^{N} D_k(t) p_{k,n} \quad 1 \leq n \leq N.$$

Similar to the development in the single queue case, let $Q_n(t)$ denote the number of customers at node n at time t, and let $Q_n(t + \Delta t)$ be the number of customers at node n at time $t + \Delta t > t$.

Also, we define $\Delta Q_n(t) = Q_n(t + \Delta t) - Q_n(t)$. Then,

$$Q_n(t) = A_n(t) - D_n(t),$$
$$\Delta Q_n(t) = \Delta A_n(t) - \Delta D_n(t),$$

where:

$$\Delta A_n(t) = A_n(t + \Delta t) - A_n(t) \quad \text{and} \quad \Delta D_n(t) = D_n(t + \Delta t) - D_n(t).$$

Let the coefficient of variation of the service time at node n be denoted by C_n, which is the ratio of the standard deviation of the service time to the mean service time; that is,

$$C_n = \mu_n \sigma_n.$$

From the results obtained in the single queue, $\Delta Q_n(t)$ is approximately normally distributed with mean and covariance from Kobayashi (1974) given by:

$$E[\Delta Q_n(t)] = \beta_n \Delta t,$$
$$\text{Cov}\{\Delta Q_n(t), \Delta Q_m(t)\} = \alpha_{nm} \Delta t,$$

$$\beta_n = \sum_{k=0}^{N} p_{k,n} \mu_k - \mu_n \equiv \lambda_n - \mu_n, \quad 1 \leq n \leq N,$$

$$\alpha_{nm} = \sum_{k=0}^{N} \mu_k (C_k^2 - 1) p_{k,n} p_{k,m} + \left\{ \mu_n C_n^2 + \sum_{k=0}^{N} p_{k,n} \mu_k \right\} \delta_{nm}$$
$$- \mu_n C_n^2 p_{n,m} - \mu_m C_m^2 p_{m,n}.$$

Using the same technique used for the single queue, let $Q(t) = [Q_n(t)]$ be an N-dimensional discrete process, and let $X(t)$ be an N-dimensional continuous-path process that is defined by the stochastic differential equation:

$$dX(t) = b\,dt + (a\,dt)^{1/2}\,Y(t),$$

where $Y(t)$ is an N-dimensional white Gaussian process, each component of which has zero mean, unit variance, and zero cross-variance; b is an N-dimensional column vector whose nth component is β_n; and a is a nonnegative definite matrix $a = [\alpha_{nm}]$ that has a unique nonnegative definite square root $a^{1/2}$. The conditional PDF of $X(t)$ is given by:

$$f_{X(t)|X(0)}(x, t|x_0) = (2\pi \det|at|)^{-1} \exp\left\{\frac{-(x - x_0 - b_t)^{\mathrm{T}} a^{-1}(x - x_0 - bt)}{2t}\right\},$$

which satisfies the multidimensional diffusion equation:

$$\frac{\partial f_{X(t)|X(0)}(x, t|x_0)}{\partial t} = \frac{1}{2}\sum_{n=1}^{N}\sum_{m=1}^{N}\alpha_{nm}\frac{\partial^2 f_{X(t)|X(0)}(x, t|x_0)}{\partial x_n \partial x_m} - \sum_{n=1}^{N}\beta_n \frac{\partial f_{X(t)|X(0)}(x, t|x_0)}{\partial x_n}.$$

The equilibrium distribution is obtained by setting the left side of the preceding equation to zero to obtain:

$$0 = \frac{1}{2}\sum_{n=1}^{N}\sum_{m=1}^{N}\alpha_{nm}\frac{\partial^2 f_X(x)}{\partial x_n \partial x_m} - \sum_{n=1}^{N}\beta_n \frac{\partial^2 f_X(x)}{\partial x_n},$$

whose solution is:

$$f_X(x) = \prod_{n=1}^{N}|\gamma_n|\exp(\gamma_n x_n),$$

where γ_n is the nth component of the N-dimensional vector:

$$\gamma = 2\alpha^{-1}\beta.$$

Following the method used in the single-queue case, we can obtain the PMF of the approximate number of customers in the network as:

$$\hat{p}_Q(q_1, q_2, \ldots, q_N) = \prod_{n=1}^{N}\hat{p}_{Q_n}(q_n),$$

$$\hat{p}_{Q_n}(m) = (1 - \eta_n)\eta_n^m,$$

$$\eta_n = \exp(\gamma_n).$$

6.3.4 Diffusion Approximation of Closed Queueing Networks

As discussed in Chapter 5, a closed queueing network is a network in which the number of customers is fixed. This is achieved either by having the customers cycle through the nodes without the opportunity to leave the system or by having a new customer enter the network as soon as a customer leaves.

One way to model and analyze the closed queueing network is to assume that when a customer enters state $N + 1$ of the open queueing network, it immediately returns to state 0, as shown in Figure 6.4.

We assume that the mean service time and variance of the service time in state 0 is zero. Let $p_{n,m}^0$ be the probability that a customer moves to node m next given that the customer is currently receiving service at node n. This probability can be obtained by noting that the customer can either move directly from node n to node m with probability $p_{n,m}$, or it can move from node n to node $N + 1$ with probability $p_{n,N+1}$ and then to node 0 with probability one and immediately to node m with probability $p_{0,m}$. Thus we have that:

$$p_{n,m}^0 = p_{n,m} + p_{n,N+1}p_{0,m} \quad 1 \leq n, m \leq N.$$

Let V_k be an N-dimensional column vector whose kth element is 1 and the mth element, $m \neq k$, is $-p_{k,m}$. That is,

$$V_k^T = [-p_{k,1}, -p_{k,2}, \ldots, -p_{k,k-1}, 1, -p_{k,k+1}, \ldots, -p_{k,N}].$$

Similarly, let W be an $N \times N$ nonnegative definite matrix whose elements are defined by:

$$w_{n,m} = \begin{cases} \displaystyle\sum_{k=0}^{N} \mu_k \{p_{k,n}(1 - p_{k,n})\} & n = m \\[4mm] \displaystyle -\sum_{k=0}^{N} \mu_k p_{k,n} p_{k,m} & n \neq m \end{cases}.$$

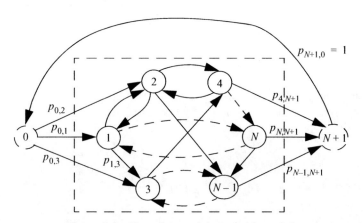

Figure 6.4 Closed queueing network model.

As shown in Kobayashi (1974), the parameters of the diffusion process are given by:

$$\beta_n^0 = \sum_{k=1}^{N} \mu_k p_{k,n}^0 - \mu_n \equiv \lambda_n - \mu_n, \quad 1 \le n \le N,$$

$$\alpha = \sum_{k=1}^{N} \mu_k C_k^2 V_k V_k^T + W.$$

Thus, if we define $\gamma = 2\alpha^{-1}\beta^0$, the equilibrium state distribution of $f_X(x)$ is given by:

$$f_X(x) = C \exp(\gamma^T x),$$

where C is a normalizing factor. Integrating $f_X(x)$ over a unit interval along each coordinate axis gives the following approximate discrete PMF:

$$\hat{p}_{Q_1 Q_2 \ldots Q_N}(m_1, m_2, \ldots, m_N) = \begin{cases} \dfrac{1}{K} \displaystyle\prod_{k=1}^{N} \eta_k^{m_k} & m_1 + m_2 + \ldots + m_N = L \\ 0 & \text{otherwise} \end{cases},$$

where L is the total number of customers in the system and $\eta_k = \exp(\gamma_k)$. The normalizing constant K is given by:

$$K = \sum_{m_1 + m_2 + \ldots + m_N = L} \prod_{k=1}^{N} \eta_k^{m_k}.$$

More information on diffusion approximation of queueing networks can be found in Glynn (1990).

6.4   PROBLEMS

6.1   Consider the two-node open network shown in Figure 6.5 in which the external arrival process has a mean interarrival time of $1/\lambda = 2$ and the

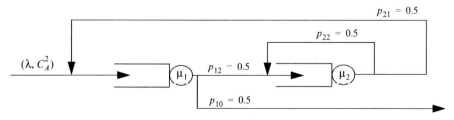

Figure 6.5   Figure for Problem 6.1.

squared coefficient of variation $C_A^2 = \sigma_A^2/(E[A])^2 = 0.94$. The service rate and the squared coefficient of variation at node 1 are given by $\mu_{X_1} = 1.1$, $C_{X_1}^2 = 0.5$, respectively; and the service rate and squared coefficient of variation at node 2 are given by $\mu_{X_2} = 1.2$, $C_{X_2}^2 = 0.8$, respectively. Obtain the joint PMF for the approximate number of customers in the network.

6.2 Consider the two-node closed network shown in Figure 6.6 in which the service rate and the squared coefficient of variation at node 1 are given by $\mu_{X_1} = 1.1$, $C_{X_1}^2 = 0.5$, respectively; and the service rate and squared coefficient of variation at node 2 are given by $\mu_{X_2} = 1.2$, $C_{X_2}^2 = 0.8$, respectively. Assuming that the total number of customers in the network is 4, obtain the joint PMF for the approximate distribution of customers in the network.

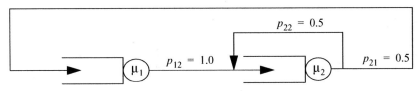

Figure 6.6 Figure for Problem 6.2.

6.3 Consider a G/G/1 queue that is empty at time $t = 0$. Assume that the interarrival times are uniformly distributed between 1 and 19 and that the service times are uniformly distributed between 2 and 17. Find the diffusion approximation for the limiting virtual waiting time of the system.

6.4 Consider an M/G/1 queue that is empty at time $t = 0$. Assume that the interarrival times are exponentially distributed with mean 10 and that the service times are uniformly distributed between 2 and 17. Find the diffusion approximation for the limiting virtual waiting time of the system.

6.5 Consider a D/G/1 queue that is empty at time $t = 0$. Assume that customers arrive every 10 time units and that the service times are uniformly distributed between 2 and 17. Find the diffusion approximation for the limiting virtual waiting time of the system.

6.6 Consider a G/G/1 queue that is empty at time $t = 0$. Assume that the interarrival times are uniformly distributed between 1 and 19 and that the service times are uniformly distributed between 2 and 17. Find the Brownian motion approximation for the limiting number of customers in the system.

6.7 Consider the Brownian motion with drift:

$$Y(t) = \mu t + \sigma B(t) + x,$$

where $Y(0) = x$, $B(t)$ is the standard Brownian motion and $b < x < a$. Let $p_a(x)$ denote the probability that $Y(t)$ hits a before b.

a. Show that:

$$\frac{1}{2}\frac{d^2 p_a(x)}{dx^2} + \mu \frac{dp_a(x)}{dx} = 0.$$

b. Deduce that $p_a(x) = \dfrac{e^{-2\mu b} - e^{-2\mu x}}{e^{-2\mu b} - e^{-2\mu a}}.$

c. What is $p_a(x)$ when $\mu = 0$?

6.8 Let $\{B(t), t \geq 0\}$ be a Brownian motion, and define the stochastic process:

$$X(0) = 0,$$
$$X(t) = tB(1/t) \quad t > 0.$$

Prove that $X(t)$ is a Brownian process.

7

ELEMENTS OF GRAPH THEORY

7.1 INTRODUCTION

Graph theory has become a primary tool for detecting numerous hidden structures in various information networks, including the Internet, social networks, and biological networks. It is a mathematical model of any system that involves a binary relation. The theory is intimately related to many branches of mathematics including group theory, matrix theory, probability, topology, and combinatorics. The intuitive appeal of graphs arises from their diagrammatic representation. They are widely used in physics, chemistry, computer science, electrical engineering, civil engineering, architecture, genetics, psychology, sociology, economics, linguistics, and operations research. In this chapter we discuss the essential aspects of graph theory that will enable us to understand Bayesian networks, Boolean networks, and random networks that are discussed in the next three chapters.

7.2 BASIC CONCEPTS

A graph is a set of points (or *vertices*) that are interconnected by a set of lines (or *edges*). For a graph G we denote the set of vertices by V and the set of edges by E and thus write $G = (V, E)$. The number of vertices in a graph is $|V|$ and the number of edges is $|E|$. The cardinality $|V|$ is called the *order* of the

Fundamentals of Stochastic Networks, First Edition. Oliver C. Ibe.
© 2011 John Wiley & Sons, Inc. Published 2011 by John Wiley & Sons, Inc.

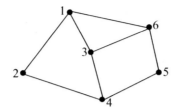

Figure 7.1 Example of a graph.

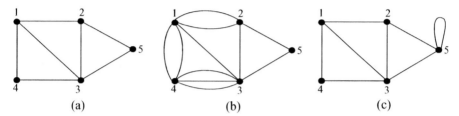

(a)   (b)   (c)

Figure 7.2 Examples of (a) a simple graph, (b) a multigraph, and (c) a graph with loop.

graph, and the cardinality $|E|$ is called the *size* of the graph. An edge is specified by the two vertices that it connects. These two vertices are the endpoints of the edge. Thus, an edge whose two endpoints are v_i and v_j is denoted by $e_{ij} = (v_i, v_j)$. If a vertex is an endpoint of an edge, the edge is sometimes said to be *incident* with the vertex. An example of a graph is shown in Figure 7.1.

In Figure 7.1, we have that:

$$V = \{1, 2, 3, 4, 5, 6\},$$
$$E = \{(1, 2), (1, 3), (1, 6), (2, 1), (2, 4), (3, 1), (3, 4), (3, 6),$$
$$(4, 2), (4, 3), (4, 5), (5, 4), (5, 6), (6, 1), (6, 3), (6.5)\}.$$

Two vertices are said to be *adjacent* if they are joined by an edge. For example, in Figure 7.1, vertices 1 and 2 are adjacent. If the edge $e_{ij} = (v_i, v_j)$ exists, we call vertex v_j a *neighbor* of vertex v_i. The set of neighbors of vertex v_i is usually denoted by $\Gamma(v_i)$.

A graph is defined to be a *simple graph* if there is at most one edge connecting any pair of vertices and an edge does not loop to connect a vertex to itself. When multiple edges are allowed between any pair of vertices, the graph is called a *multigraph*. Examples of a simple graph, a multigraph, and a graph with loop are shown in Figure 7.2.

The *degree* (or *valency*) of a vertex v, which is denoted by $d(v)$, is the number of edges that are incident with v. For example, in Figure 7.2a, $d(3) = 4$ and $d(4) = 2$. If a vertex v has $d(v) = 0$, then v is said to be *isolated*. Similarly, a vertex of degree one (i.e., $d(v) = 1$) is called a *pendant vertex* or an *end vertex*. It can be shown that:

$$\sum_{v \in V} d(v) = 2|E|,$$

where $|E|$ is the number of edges in the graph.

A *regular graph* is a graph in which each vertex has the same degree. A *k-regular* graph is one in which the degree of each vertex is k.

7.2.1 Subgraphs and Cliques

A *subgraph* of G is a graph H such that the vertices of H, denoted by $V(H)$, are a subset of V (i.e., $V(H) \subseteq V$) and the edges of H, denoted by $E(H)$, are a subset of E (i.e., $E(H) \subseteq E$), and the endpoints of an edge $e \in E(H)$ are the same as its endpoints in G. For example, Figure 7.3 is a subgraph of the graph in Figure 7.2a. Note that vertex 2 is an example of a pendant vertex.

A *complete graph* K_n on n vertices is the simple graph that has all $\binom{n}{2}$ possible edges. The *density* of a graph is the ratio of the number of edges in the graph to the number of edges in a complete graph with the same number of vertices. Thus, the density of the graph $G = (V, E)$ is $2|E|/|V|\{|V| - 1\}$. Figure 7.4 illustrates examples of complete graphs.

A *clique* of a graph G is a single node or a complete subgraph of G. That is, a clique is a subgraph of G in which every vertex is a neighbor of all other vertices in the subgraph. Figure 7.5 shows examples of cliques.

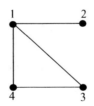

Figure 7.3   Example of a subgraph of Figure 7.2a.

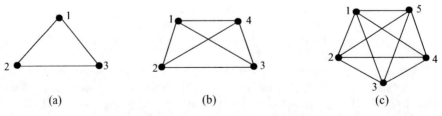

(a)   (b)   (c)

Figure 7.4   Examples of complete graphs. (a) K_3 complete graph; (b) K_4 complete graph; (c) K_5 complete graph.

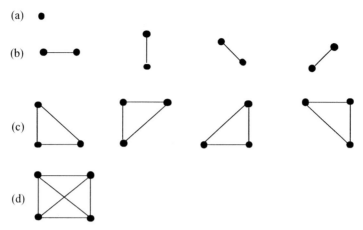

Figure 7.5 Examples of cliques. (a) One-node clique; (b) two-node cliques; (c) three-node cliques; (d) four-node cliques.

7.2.2 Adjacency Matrix

Another way to describe a graph is in terms of the *adjacency matrix* $A(x, y)$, which has a value of 1 in its (x, y) cell if x and y are neighbors and zero otherwise, for all $x, y \in V$. Then the degree of vertex x is given by:

$$d(x) = \sum_y A(x, y).$$

For example, the adjacency matrix of the graph in Figure 7.1 is as follows:

$$A = \begin{bmatrix} 0 & 1 & 1 & 0 & 0 & 1 \\ 1 & 0 & 0 & 1 & 0 & 0 \\ 1 & 0 & 0 & 1 & 0 & 1 \\ 0 & 1 & 1 & 0 & 1 & 0 \\ 0 & 0 & 0 & 1 & 0 & 1 \\ 1 & 0 & 1 & 0 & 1 & 0 \end{bmatrix}.$$

7.2.3 Directed Graphs

What we have described so far is often referred to as an *undirected graph*. Graphs are often used to model relationships. When there is a special association in these relationships, the undirected graph does not convey this information; a *directed graph* is required. A directed graph (or *digraph*) is a graph in which an edge consists of an ordered vertex pair, giving it a direction from one vertex to the other. Generally in a digraph the edge (a, b) has a direction from vertex a to vertex b, which is indicated by an arrow in the direction from a to

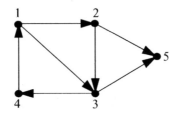

Figure 7.6 Example of a digraph.

b. Figure 7.6 illustrates a simple digraph. When the directions are ignored, we obtain the underlying undirected graph shown earlier in Figure 7.2a.

In this case the adjacency matrix becomes the following:

$$A = \begin{bmatrix} 0 & 1 & 1 & 0 & 0 \\ 0 & 0 & 1 & 0 & 1 \\ 0 & 0 & 0 & 1 & 1 \\ 1 & 0 & 0 & 0 & 0 \\ 0 & 0 & 0 & 0 & 0 \end{bmatrix}.$$

In a directed graph, the number of edges that are directed into a vertex is called the *indegree* of the vertex, and the number of edges that are directed out of a vertex is called the *outdegree* of the vertex. The sum of the indegree and the outdegree of a vertex is the degree of the vertex.

7.2.4 Weighted Graphs

In some applications, the edge between two vertices has a specific meaning. For example, if the vertices represent towns, the edges can be used to represent the distances between adjacent towns. Similarly, if a graph is used to model a transportation network, the edges may represent the capacity of the link between two vertices in the network. When the edges are associated with this special function, the graph will have a nonnegative number called a *weight* that is assigned to each edge. Thus, we consider a connected graph with n vertices labeled $\{1, 2, \ldots, n\}$ with weight $w_{ij} \geq 0$ on the edge (i, j). If edge (i, j) does not exist, we set $w_{ij} = 0$. We assume that the graph is undirected so that $w_{ij} = w_{ji}$. Such a graph is called a *weighted graph*.

The total weight of the edges emanating from vertex i, w_i, is given by:

$$w_i = \sum_j w_{ij},$$

and the sum of the weights of all edges is:

$$w = \sum_{i,j:j>i} w_{ij},$$

where the inequality in the summation is used to avoid double counting.

7.3 CONNECTED GRAPHS

A *walk* (or *chain*) in a graph is an alternating sequence $v_0, e_1, v_1, e_2, \ldots, v_{k-1}$, e_k, v_k of vertices v_i, which are not necessarily distinct, and edges e_i such that the endpoints of e_i are v_{i-1} and v_i, $i = 1, \ldots, k$. A *path* is a walk in which the vertices are distinct. For example, in Figure 7.2a, the path $\{1, 3, 5\}$ connects vertices 1 and 5.

A walk in which all the edges are distinct is called a *trail*. Note that a path is a trail in which the vertices are distinct, except possibly the first and last vertices. A *closed walk* is a walk in which the first and last vertices are the same; that is, it is a walk between a vertex and itself. A closed walk with distinct edges is called a *circuit*, and a closed walk with distinct vertices is called a *cycle*. Alternatively, a cycle is a circuit with no repeated intermediate vertices.

When a walk can be found between every pair of distinct vertices, we say that the graph is a *connected graph*. A graph that is not connected can be decomposed into two or more connected subgraphs, each pair of which has no vertex in common. That is, a *disconnected graph* is the union of two or more disjoint connected subgraphs. When a path can be found between every pair of distinct vertices, we say that the graph is a *strongly connected graph*. The *distance* between two vertices is the number of edges connecting them. The *geodesic distance* between two vertices is the length of the shortest path between them.

7.4 CUT SETS, BRIDGES, AND CUT VERTICES

Let F be the set of edges in the graph $G = (V, E)$. The graph that is obtained by deleting from G all the edges in F is denoted by $G - F$. If $G - F$ has more than one component, the set F is called a *disconnecting set*. If F is a disconnecting set that consists of only one edge, the edge is called a *bridge* or a *cut edge*. A graph in which every disconnecting set in it has at least k edges is called a *k-edge connected* graph. The minimum size of a disconnecting set in G is called the *edge-connectivity number* $\lambda(G)$ of the graph. A disconnecting set F in which no proper subset of F is a disconnecting set is called a *cut set*. Figure 7.7 is an example of a disconnecting set where $F = \{(1,2), (1,3), (4,3)\}$. Note that F is a cut set since no proper subset of F is a disconnecting set.

Let U be a set of vertices in $G = (V, E)$. The graph obtained by deleting from G all the vertices in U together with all the edges that are incident with

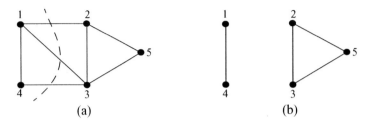

Figure 7.7   Example of a disconnecting set. (a) G; (b) $G - F$.

these vertices in U is denoted by $G - U$. If $G - U$ has more than one component, U is called a *separating set* or a *vertex cut*. If a separating set consists of only one vertex u, then u is called a *cut vertex*. The minimum size of a separating set in G is called the *connectivity number* $\kappa(G)$. It is well known that $\kappa(G) \leq \lambda(G)$.

7.5   EULER GRAPHS

An *Euler trail* is a trail that contains every edge of G, and an *Euler circuit* is a circuit that contains every edge of G. In other words, in an Euler trail and an Euler circuit every edge is traversed exactly once; the difference is that an Euler circuit is a closed trail. A connected graph that contains an Euler circuit is called an *Euler graph*, and a connected graph that contains an Euler trail is called a *semi-Eulerian graph*. An undirected graph possesses an Euler circuit if and only if it is connected and its vertices are all of even degree. Similarly, a directed graph possesses an Euler circuit if and only if it is connected and the indegree of every vertex is equal to its outdegree.

7.6   HAMILTONIAN GRAPHS

A *Hamiltonian path* between two vertices is a path that passes through every vertex of the graph. A *Hamiltonian cycle* is a closed Hamiltonian path. A *Hamiltonian graph* is a graph that has a Hamiltonian cycle. Unfortunately, unlike the Euler graph, there is no elegant way to characterize a Hamiltonian graph. The following two theorems apply to undirected graphs:

Theorem 7.1 (Ore's Theorem): If G is a simple graph with n vertices, where $n \geq 3$, and if for every pair of nonadjacent vertices u and v, $d(u) + d(v) \geq n$, where $d(x)$ is the degree of vertex x, then G is a Hamiltonian graph.

Theorem 7.2 (Dirac's Theorem): If G is a simple graph with n vertices, where $n \geq 3$, and if $d(v) \geq n/2$ for every vertex v, then G is a Hamiltonian graph.

Similarly, for directed graphs we have the following theorems:

Theorem 7.3: A digraph with n vertices is Hamiltonian if both the indegree and outdegree of every vertex is at least $n/2$.

Theorem 7.4: A digraph with n vertices has a directed Hamiltonian path if any of the following conditions holds:

1. The sum of the degrees of every pair of nonadjacent vertices is at least $2n - 3$.
2. The degree of each vertex is at least $n - 1$.
3. Both the indgree and outdegree of each vertex are at least $(n - 1)/2$.

7.7   TREES AND FORESTS

A *tree* is a connected graph G that has no circuit. This implies that G cannot be a multigraph. A collection of disjoint trees is called a *forest*. An example of a tree is shown in Figure 7.8.

Vertex v_0 is called the *root* of the tree. The root has three branches that terminate at vertices v_1, v_2, v_3 that are the *children* or *offspring* of v_0. Each child may have its own children. Each branch has a last *terminal edge* to a *terminal vertex* (also called a *leaf vertex*) that has no further branches. For example, vertices v_4 and v_{15} are terminal vertices. A tree with n vertices has $n - 1$ edges.

Figure 7.9 shows an example of a forest that consists of three trees (or components). A forest with k components and n vertices has $n - k$ edges. There are 13 edges in Figure 7.9.

A *spanning tree* of a graph is a tree of the graph that contains all the vertices of the graph. A graph is connected if and only if it has a spanning tree. For a graph with k components, a spanning tree can be generated for each component, and the k spanning trees are called a *spanning forest*. Figure 7.10 shows a two-component graph with an associated spanning forest.

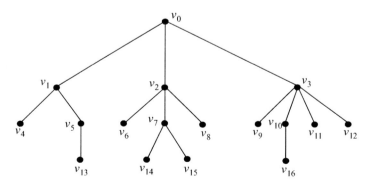

Figure 7.8   Example of a tree.

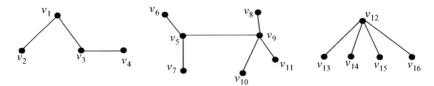

Figure 7.9   Example of a forest.

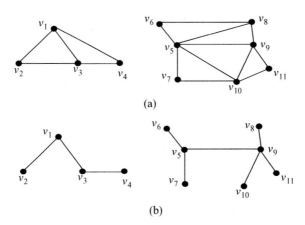

Figure 7.10   Example of a spanning forest. (a) A two-component graph; (b) a two-component spanning forest.

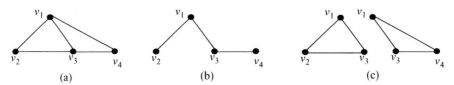

Figure 7.11   Example of the fundamental cycle of a spanning tree. (a) Graph G; (b) a spanning tree T; (c) fundamental cycle of T.

Let T be a spanning tree of a graph G with n nodes and m edges. Suppose we add to T an edge in $G - T$, which is an edge of G that is not in T. This will result in a cycle. The set of all cycles obtained by separately adding edges in $G - T$ is called the *fundamental cycle* of T. Since T is not necessarily unique, we may refer to its fundamental cycle as a fundamental cycle of G. The number of cycles in the fundamental cycle of T is $m - (n - 1) = m - n + 1$; this is referred to as the *cycle rank* of G. For example, a spanning tree of a graph G and its fundamental cycle are shown in Figure 7.11. In this case, the cycle rank of G is $5 - 4 + 1 = 2$.

The fundamental cycle is popularly used in solving electric network problems.

7.8   MINIMUM WEIGHT SPANNING TREES

Let T be a spanning tree in a weighted graph and let $w(T)$ be the sum of the weights of all the edges in T. That is,

$$w(T) = \sum_{(i,j)\in T} w_{ij}.$$

A spanning tree T in a weighted graph G is called a minimum weight spanning tree if $w(T) \le w(T_k)$ for every spanning tree T_k in G. One of the algorithms for obtaining a minimum weight spanning tree is the Kruskal's algorithm, which is a greedy algorithm that operates as follows:

Step 1: Arrange the edges in the graph in a nondecreasing order of their weights as a list L and set T to be an empty set.

Step 2: Select the first edge from L and include it in T.

Step 3: If every edge in L has been examined for possible inclusion in T, stop and declare that G is unconnected. Otherwise, take the first unexamined edge in L and see if it forms a cycle in T. If it does not form a cycle, select it for inclusion in T and go to step 4; otherwise, mark the edge as examined and repeat step 3.

Step 4: If T has $n - 1$ edges, stop; otherwise, go to step 3.

Example 7.1: Consider the graph shown in Figure 7.12. Here, the number of nodes is $n = 4$. Thus, the cardinality of T is 3.

The list L is given by $L = \{(1, 2), (1, 3), (2, 3), (1, 4), (3, 4)\}$ and initially we have $T = \{\ \}$. Thus, we select the first edge $(1, 2)$ and have $T = \{(1, 2)\}$. We next select the edge $(1, 3)$ in L and include it in T since including it does not result in a cycle. We update T: $T = \{(1,2), (1.3)\}$. The next edge in L is $(2, 3)$. However, including it in T results in a cycle, so it is marked and not included. The next

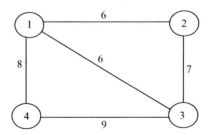

Figure 7.12   Example of a weighted graph.

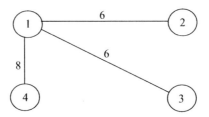

Figure 7.13   Minimum weight spanning tree of graph in Figure 7.12.

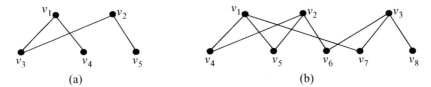

Figure 7.14   Examples of a bipartite graph.

node in L is $(1, 4)$. Since including it in T does not result in a cycle, we include it and update T: $T = \{(1, 2), (1.3), (1, 4)\}$. Since the number of edges in T is 3, we stop. Thus, the minimum weight spanning tree is shown in Figure 7.13 with $w(T) = 20$.

7.9   BIPARTITE GRAPHS AND MATCHINGS

Consider a graph $G = (V, E)$ whose vertices can be partitioned into two subsets V_1 and V_2. If it is possible that every edge of G connects a vertex in V_1 to a vertex in V_2, then G is called a *bipartite graph*. Thus, a bipartite graph is a graph in which it is possible to partition the vertex set into two subsets V_1 and V_2 such that every edge of the graph joins a vertex of V_1 with a vertex of V_2, and no vertex is adjacent to another vertex in its own set. Figure 7.14 shows examples of a bipartite graph. In (a), $V_1 = \{v_1, v_2\}$ and $V_2 = \{v_3, v_4, v_5\}$. Similarly, in (b) we have that $V_1 = \{v_1, v_2, v_3\}$ and $V_2 = \{v_4, v_5, v_6, v_7, v_8\}$.

If every vertex of V_1 is connected to every vertex of V_2, the graph is said to be a complete bipartite graph, which is denoted by $K_{m, n}$, where $|V_1| = m$ and $|V_2| = n$. Figure 7.15 shows a $K_{2,3}$ bipartite graph and a $K_{3,5}$ bipartite graph.

The notion of a bipartite graph can be generalized to partitioning the vertex set into s subsets: $V_1, V_2, \ldots, V_s$, where $s > 2$. An s-*partite graph* is a graph in which the vertex set is partitioned into s subsets and all the edges of the graph are between vertices in different subsets and no vertices in the same subset are adjacent. A *complete s-partite graph* is an s-partite graph in which all pairs of vertices belonging to different subsets are adjacent.

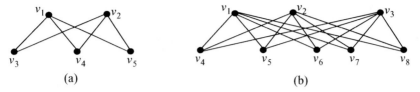

(a)   (b)

Figure 7.15   Examples of complete bipartite graphs. (a) $K_{2,3}$ bipartite graph; (b) $K_{3,5}$ bipartite graph.

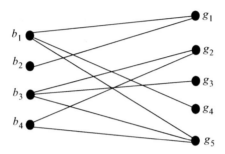

Figure 7.16   Bipartite graph for the marriage problem.

A set of edges M in a graph G is called a *matching* (or an *independent edge set*) in G if no two edges in M have a vertex in common. A matching with the largest possible number of edges is called a *maximum matching*. If every vertex of G is the endpoint of an edge in a matching M, then M is said to be a *perfect matching*. Many problems can be formulated as a maximum matching problem. One example is the so-called *marriage problem*: If there is a finite set of boys, each of whom knows several girls, under what conditions can all the boys marry the girls in such a way that each boy marries a girl he knows? Such a problem can be represented by a bipartite graph $G = G(V_1, V_2)$, in which V_1 represents the set of boys and V_2 represents the set of girls and an edge exists between a node in V_1 and a node in V_2 if the boy and girl represented by those nodes know each other. Hall's theorem states the necessary and sufficient condition for the solution of the marriage problem:

Theorem 7.5 (Hall's Theorem): A necessary and sufficient condition for a solution to the marriage problem is that each of k boys collectively knows at least k girls, where $1 \le k \le m$ and m is the total number of boys.

For example, consider the case where the number of boys is 4 and the number of girls is 5 and the friendships are as shown in Figure 7.16. As can be seen, Hall's theorem is satisfied, which means that a solution can be found for the problem. One such solution is $\{(b_1, g_4), (b_2, g_1), (b_3, g_3), (b_4, g_2)\}$.

Assume that each edge e of a bipartite graph is assigned a nonnegative weight w_e. The problem of finding a matching M in G such that the sum of all the weights of the edges in M is as small as possible is called the *optimal assignment problem*. In its most general form, the assignment problem can be stated as follows: There are m agents and n tasks. Any agent can be assigned to perform any task, and the cost of assigning agent i to task k is c_{ik}. It is required to perform all tasks by assigning exactly one agent to each task in such a way that the total cost of the assignment is minimized. The agents constitute the vertex set V_1 while the tasks constitute the vertex set V_2 of the bipartite graph $G = G(V_1, V_2)$. This problem can be solved as a linear programming problem using such algorithms as the simplex method, as discussed in texts on combinatorial optimization such as Papadimitriou and Steiglitz (1982). Another popular method of solution is the so-called *Hungarian method*. We discuss the matrix form of the Hungarian method next.

7.9.1 The Hungarian Algorithm

The algorithm is defined here for minimizing the total cost of assignment. The agents form the rows of the matrix and the tasks form the columns with the entry c_{ij} being the cost of using agent i to perform task j. If the matrix is not square, we make it square by adding row(s) or column(s) of zeroes when necessary. A maximum assignment can be converted into a minimum assignment by replacing each entry c_{ij} with $C - c_{ij}$, where C is the maximum value in the assignment matrix. The Hungarian algorithm proceeds in the following steps:

1. Subtract the minimum number in each row from each entry in the entire row.
2. Subtract the minimum number in each column from each entry in the entire column.
3. Cover all zeroes in the matrix with as few lines (horizontal and/or vertical only) as possible. Let k be the number of lines and n the size of the matrix.
   - If $k < n$, let m be the minimum uncovered number. Subtract m from every uncovered number and add m to every number covered by two lines. Go back to step 3.
   - If $k = n$, go to step 4.
4. Starting with the top row, work your way downwards as you make assignments. An assignment can be (uniquely) made when there is exactly one zero in a row. Once an assignment is made, delete that row and column from the matrix.

Example 7.2: Consider the assignment problem in Figure 7.17 where the weight of each edge is the cost of agent Al, Bob, Chris, or Dan performing task W or X or Y or Z.

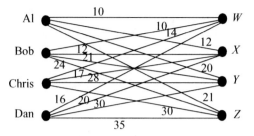

Figure 7.17 Bipartite graph for the assignment problem.

The solution begins with the problem displayed in a matrix form as follows:

$$M = \begin{bmatrix} 10 & 12 & 20 & 21 \\ 10 & 12 & 21 & 24 \\ 14 & 17 & 28 & 30 \\ 16 & 20 & 30 & 35 \end{bmatrix}.$$

Step 1: We subtract the smallest entry of each row from every entry in the row and obtain the following matrix:

$$M_1 = \begin{bmatrix} 0 & 2 & 10 & 11 \\ 0 & 2 & 10 & 14 \\ 0 & 3 & 14 & 16 \\ 0 & 4 & 14 & 19 \end{bmatrix}.$$

Step 2: We subtract the smallest entry of each column from every entry in the column to obtain the following matrix:

$$M_2 = \begin{bmatrix} 0 & 0 & 0 & 0 \\ 0 & 0 & 0 & 3 \\ 0 & 1 & 4 & 5 \\ 0 & 2 & 4 & 8 \end{bmatrix}.$$

Step 3: We cover the zeroes with as few horizontal and vertical lines as possible. We can cover them with a vertical line in the first column and two horizontal lines in the first and second columns as follows:

$$M_2 = \begin{bmatrix} \cancel{0} & \cancel{0} & \cancel{0} & \cancel{0} \\ \cancel{0} & \cancel{0} & \cancel{0} & \cancel{3} \\ 0 & 1 & 4 & 5 \\ 0 & 2 & 4 & 8 \end{bmatrix}.$$

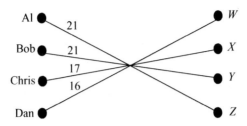

Figure 7.18   Solution of the assignment problem.

Since the number of lines is less than 4, we identify 1 as the smallest uncovered entry and subtract it from every uncovered entry and add it to the first-row, first-column entry and the second-row, first-column entry that have two lines and obtain the following matrix:

$$M_3 = \begin{bmatrix} 1 & 0 & 0 & 0 \\ 1 & 0 & 0 & 2 \\ 0 & 0 & 3 & 4 \\ 0 & 1 & 3 & 7 \end{bmatrix}.$$

We cover the zeroes with as few horizontal and vertical lines as possible to obtain the following:

$$M_3 = \begin{bmatrix} 1 & 0 & 0 & 0 \\ 1 & 0 & 0 & 2 \\ 0 & 0 & 3 & 4 \\ 0 & 1 & 3 & 7 \end{bmatrix}.$$

Since we have four lines, an assignment can be made. We start with the fourth row with only one zero and assign Dan to task W, which eliminates column 1. We move up to the third row and assign Chris to task X, which eliminates columns 1 and 2. We next assign Bob to task Y and assign Al to task Z. Thus the optimal assignment is shown in Figure 7.18 with a total cost of 75.

7.10   INDEPENDENT SET, DOMINATION, AND COVERING

A set of vertices of a graph G is called an *independent set* or an *internally stable set* if no two vertices in the set are adjacent. An independent set that is not contained in any other independent set is called a *maximal independent set*. The cardinality of the set of vertices of G that is the largest independent set is called the *independence number* of G and is denoted by $\alpha(G)$. Thus, the independence number is the cardinality of the *maximum independence set*.

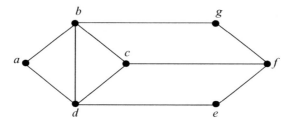

Figure 7.19   Graph for independent set example.

Example 7.3: Consider the graph in Figure 7.19. The vertex sets $\{a, c, g\}$, $\{b, f\}$, and $\{a, c, e, g\}$ are examples of independent sets. The independent sets $\{b, f\}$ and $\{a, c, e, g\}$ are maximal independent sets while $\{a, c, e, g\}$ is the maximum independent set, which means that the independence number of the graph is 4, the cardinality of $\{a, c, e, g\}$.

A set $D \in V$ of the vertices of a graph G is defined to be a *dominating set* or an *externally stable set* of G if every node in $V - D$ has at least one neighbor in D. For example, the vertex sets $\{b, f\}$, $\{a, f\}$, and $\{a, c, f\}$ of the graph in Figure 7.19 are dominating sets of the graph. Note that a dominating set need not be an independent set as the dominating set $\{a, c, f\}$ illustrates. A dominating set is called a *minimal dominating set* if it is not contained in any other dominating set. A dominating set that has the smallest cardinality is called a *minimum dominating set* of G, and its cardinality is called the *domination number* $\beta(G)$. For the graph in Figure 7.19, the domination number is 2. Dominating sets are useful in determining the minimum number of monitoring and surveillance systems that can be deployed to ensure full coverage of a given area.

A set C of edges of a graph G is said to *cover* G if every vertex in G is a terminal vertex of at least one edge in C. A set of edges that covers a graph G is called a *covering* of G. Since C can be considered as dominating the vertices of G, a covering with the smallest cardinality (i.e., the *minimum covering*) can be called the *minimum dominating edge set* of G. For example, the link sets $\{(a, b), (c, d), (e, f), (b, g)\}$, $\{(a, b), (c, f), (d, e), (f, g)\}$, and $\{(a, b), (b, g), (c, d), (c, f), (d, e)\}$ are examples of coverings of the graph. A covering from which no edge can be removed without destroying its ability to cover the graph is called a *minimal covering*. The number of edges in a minimal graph of the smallest size is called the *covering number* $\gamma(G)$ of the graph. Note that every pendant edge in a graph must be included in every covering of the graph. Also, no minimal covering can contain a circuit, which means that a graph with n vertices can contain no more than $n - 1$ edges. Finally, the covering of an n-vertex graph will have a covering number of at least $\lceil n/2 \rceil$, where $\lceil x \rceil$ is the smallest integer that is greater than x. For example, the covering number of the graph in Figure 7.19 is 4, which is $\lceil 7/2 \rceil$. As in the case of dominating sets, covering can be used to determine the minimum number of monitoring systems that can be deployed to ensure that a given set of points can be monitored.

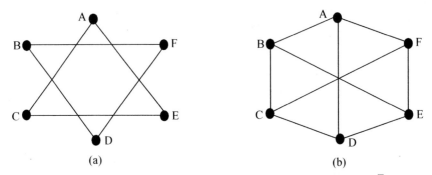

Figure 7.20 Example of (a) a graph G and (b) its complement $\bar{G}$.

7.11 COMPLEMENT OF A GRAPH

Let G be a graph with n vertices. We define the complement $\bar{G}$ of G as the graph with the same vertex set as G and where two vertices x and y that are not adjacent in G are adjacent in $\bar{G}$. Alternatively, $\bar{G}$ is obtained from the complete graph K_n by deleting all the edges of G. Figure 7.20 shows an example of a graph and its complement.

7.12 ISOMORPHIC GRAPHS

Two graphs G_1 and G_2 are said to be isomorphic if there a one-to-one mapping from the vertices of G_1 to the vertices of G_2 that ensures that the number of edges joining any two vertices of G_1 is equal to the number of edges joining the corresponding vertices of G_2. If we use the notation $\phi(x) = y$ to indicate that vertex x in G_1 is mapped into vertex y in G_2, then the two graphs shown in Figure 7.21 are isomorphic under the correspondence $\phi(a) = 1$, $\phi(b) = 4$, $\phi(c) = 2$, $\phi(d) = 5$, $\phi(e) = 3$, and $\phi(f) = 6$.

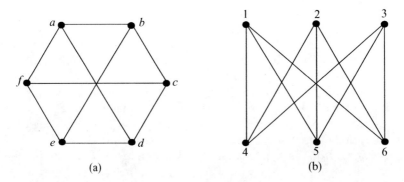

Figure 7.21 Example of isomorphic graphs. (a) G_1 isomorphic graph; (b) G_2 isomorphic graph.

One advantage of isomorphism is that if two graphs are isomorphic, then they are identical on all graph theoretic properties. For example, since G_1 and G_2 in Figure 7.21 are isomorphic and G_2 is a bipartite graph, the analysis of G_1 is simplified by the fact that there is a rich set of properties associated with its corresponding bipartite graph G_2. Thus, answers to all questions related to G_1 can be obtained from G_2, to which it is isomorphic.

7.13   PLANAR GRAPHS

A planar graph is a graph that can be drawn in the plane in such a manner that no two edges of the graph intersect except possibly at a vertex to which they are both incident. The problem of determining if a graph is planar has application in printed circuits where it is desired that lines do not cross. It is also used to deal with the well-known problem of determining whether it is possible to connect three houses a, b, and c to three utilities d, e, and f in such a way that no two connecting pipelines meet except at their initial and terminal points.

Consider the graph in Figure 7.22a. Even though it has edges crossing, it is planar because it is equivalent (or isomorphic) to the graph in Figure 7.22b that has no crossing edges.

We illustrate some of the properties of planar graph through the following theorems.

Theorem 7.6: A graph is planar if and only if it can be mapped onto the surface of a sphere such that no two edges meet one another except at the vertex or vertices with which the edges are incident.

Theorem 7.7: A linear planar graph that has no loops can be drawn on a plane with straight line segments as edges.

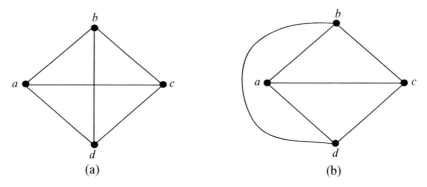

Figure 7.22   Examples of a planar graph.

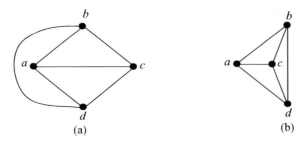

Figure 7.23   Equivalent planar graphs.

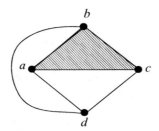

Figure 7.24   Region of a planar graph.

Theorem 7.7 has some interesting results. It states that the planar graph in Figure 7.23a can be redrawn as in Figure 7.23b in which all the edges are straight-line segments.

7.13.1   Euler's Formula for Planar Graphs

A *region* of a planar graph is an area of the plane that is bounded by edges and contains no edge and no vertex. For example, the shaded area in Figure 7.24 is a region of the graph.

A planar graph has multiple *interior regions* and one *exterior region* with an infinite area. The boundary of an interior region is the subgraph formed by the vertices and edges that encompass the region. The degree of a region is the number of edges in a closed walk that encloses it. For example, the degree of the shaded region in Figure 7.24 is 3. The following theorem gives the Euler's formula for planar graphs:

Theorem 7.8 (Euler's Formula): If a connected planar graph has v vertices, e edges, and r regions, then $v - e + r = 2$.

Theorem 7.9: If G is a connected simple planar graph with n vertices, where $n \geq 3$, and m edges, then $m \leq 3n - 6$. If, in addition, G has no triangles, then $m \leq 2n - 4$.

Theorem 7.10: K_5 and $K_{3,3}$ are nonplanar.

The proof of this theorem lies in the fact that for a K_5 graph, $n = 5$ and $m = 10$, and the graph contains triangles. Thus from Theorem 7.9 we have that $3n - 6 = 9$, which is less than 10. Also, $K_{3,3}$ has $n = 6$ and $m = 9$. Since $K_{3,3}$ has no triangle, we have that $2n - 4 = 8$, which is less than 9.

7.14   GRAPH COLORING

By coloring a graph we mean to paint the vertices of the graph with one or more distinct colors. A graph G is said to be *k-colorable* if it is possible to assign one of k colors to each vertex such that adjacent vertices have different colors. If G is k-colorable but not $(k - 1)$-colorable, we say that G is a *k-chromatic* graph and that its *chromatic number* $\chi(G)$ is k. For a bipartite graph $\chi(G) = 2$, and for a complete graph K_n, $\chi(G) = n$.

While no efficient procedure is known for finding the chromatic number of an arbitrary graph, some well-known bounds have been found. Let $\Delta(G)$ denote the maximum of the degrees of the vertices of a graph G. The following theorems establish an upper bound for $\chi(G)$.

Theorem 7.11: For any graph G, $\chi(G) \leq \Delta(G) + 1$, with equality if G is a complete graph.

Theorem 7.12 (Brooks' Theorem): If G is a simple connected graph that is not a complete graph, and if the largest vertex-degree of G is $\Delta(G)$, where $\Delta(G) \geq 3$, then $\chi(G) \leq \Delta(G)$; that is, G is $\Delta(G)$-colorable.

7.14.1   Edge Coloring

Coloring can also be applied to edges. A graph G is said to be *k-edge colorable* if it is possible to assign k colors to its edges so that no two adjacent edges have the same color, where two edges are defined to be adjacent if they are incident with the same vertex. If G is k-edge colorable but not $(k - 1)$-edge colorable, it is said to be a *k-edge chromatic graph*, and its *chromatic index* $\chi'(G)$ is k. As in the case of the vertex coloring, there are a few theorems associated with edge coloring.

Theorem 7.13: The chromatic index of a complete bipartite graph $K_{m,n}$ is $\chi'(G) = \max\{m, n\}$.

Theorem 7.14: The chromatic index of a complete graph K_n is:

$$\chi'(K_n) = \begin{cases} n & n \text{ odd} \\ n-1 & n \text{ even} \end{cases}.$$

Theorem 7.15 (Vizing's Theorem): If G is a simple graph, then $\Delta(G) \leq \chi'(G) \leq \Delta(G) + 1$. Thus, in this case, $\chi'(G)$ is either $\Delta(G)$ or $\Delta(G) + 1$.

7.14.2 The Four-Color Problem

A famous problem in the theory of graphs is to determine the smallest number of distinct colors that are sufficient to color a planar graph. The following theorem shows that five colors are always sufficient:

Theorem 7.16: Any planar graph is 5-colorable.

A question in graph theory that remained unanswered for a long time is the following: Is a planar graph 4-colorable? This is the famous four-color problem that was conjectured in 1852 when Francis Guthrie was trying to color the map of the counties of England (see Saaty 1972). The conjecture has now been proved to be true by Appel and Haken (1976, 1977).

7.15   RANDOM GRAPHS

A random graph is a finite set of vertices where edges connect vertex pairs in a random manner. Random graphs are widely used in social networks where the vertices may represent people (or "actors") and an edge exists between two vertices if there is a social tie between the actors represented by the vertices. The importance of random graphs in this case arises from the fact that these relationships are not fixed; they are random events that can change with time and circumstances.

We consider three classes of random graphs: Bernoulli random graphs, geometric random graphs, and Markov random graphs.

7.15.1   Bernoulli Random Graphs

Bernoulli random graphs (also called the Erdos–Renyi graphs) are the earliest random graphs proposed by Erdos and Renyi (1959, 1960) and are the simplest to analyze. A graph $G(N, p)$ is called a Bernoulli random graph if it has N vertices and a vertex is connected to another vertex with probability p, where $0 < p \leq 1$. It is assumed that the connections occur independently. Thus, if we assume that the edges are independent and that self-loops are forbidden, the probability that a random graph has m edges is given by:

$$\binom{M}{m} p^m (1-p)^{M-m},$$

where $M = \binom{N}{2} = N(N-1)/2$ is the total possible number of edges.

Consider a vertex in the graph. It is connected with equal probability to each of the $N-1$ other vertices, which means that if K denotes the number of vertices that it is connected with, the probability mass function (PMF) of K is given by:

$$p_K(k) = P[K = k] = \binom{N-1}{k} p^k (1-p)^{N-1-k} \quad k = 0, 1, \ldots, N-1.$$

Thus, the expected value of K is $E[K] = (N-1)p$, which is the average degree of a vertex. Observe that the average degree is not bounded as $N \to \infty$, which is a limitation of the model. If we hold the mean $(N-1)p = \beta$ so that $p = \beta/(N-1)$, then the PMF becomes:

$$p_K(k) = \binom{N-1}{k} \left[\frac{\beta}{N-1} \right]^k \left[1 - \frac{\beta}{N-1} \right]^{N-1-k}$$

$$\lim_{N \to \infty} p_K(k) = \frac{\beta^k}{k!} e^{-\beta}.$$

We recognize the quantity in the last equation as the Poisson distribution. A random graph in which each link forms independently with equal probability and the degree of each vertex has the Poisson distribution is called a *Poisson random graph*.

7.15.1.1 *Phase Transition* A component of a graph is a subset of vertices each of which is reachable from the others by some path through the graph. In the Bernoulli random graph, almost all vertices are disconnected at small values of p. As p increases the vertices increasingly form clusters, where a cluster is a subgraph such that each vertex in the subgraph can be reached from any other vertex in the subgraph via a path that lies entirely within the subgraph, and there is no edge between a vertex in the subgraph and a vertex outside of the subgraph. Generally, below $p = 1/N$, the graph consists of only small disconnected clusters whose size is $O(\log N)$, independent of p, and the average number of edges per vertex is less than one. However, above $p = 1/N$, a phase transition occurs such that a single cluster or giant component emerges that contains a finite fraction γ of the total number of vertices; that is, the expected number of vertices in the "giant" component is γN. Other clusters become vanishingly small in comparison to the giant component.

7.15.2 Geometric Random Graphs

A geometric random graph $G(N, r)$ is a graph whose vertices correspond to uniformly randomly distributed points in a metric space with distance measure δ and there is an edge between two vertices if the distance between the corresponding points in the metric space is at most r. The parameter r is called the radius of the graph. That is, the edge set is obtained as follows:

$$E = \{(u, v)|[(u, v \in V) \cap (0 < \delta(u, v) \le r)]\}.$$

If the two-dimensional unit square is our metric space, then the Euclidean distance can be the distance measure. In this case, consider a two-dimensional space whose area is A. If the quantity $(N - 1)\pi r^2 / A = \lambda$ is constant, the probability that a node has a degree k is given by:

$$p_K(k) = \binom{N-1}{k} \left[\frac{\pi r^2}{A} \right]^k \left[1 - \frac{\pi r^2}{A} \right]^{N-1-k}$$

$$\lim_{\substack{N \to \infty \\ A \to \infty}} p_K(k) = \frac{\lambda^k}{k!} e^{-\lambda}.$$

That is, the limiting distribution of the degree of a node also follows the Poisson distribution with the mean degree being λ.

7.15.3 Markov Random Graph

Markov graphs generalize the Bernoulli random graph. They allow the probability that a given edge is formed to be dependent on whether or not neighboring edges are formed. Specifying these interdependencies requires special structure. For example, making one edge dependent on a second edge and the second edge dependent on the third implies that some interdependencies between the first edge and the third edge exist; that is, the interdependencies can be transitive. A Markov graph attempts to create a model that has an element of transitivity.

Frank and Strauss (1986) applied the Hammersley–Clifford theorem (see Besag 1974) to obtain a characterization of the probability function of a realization of a random undirected graph G with dependence graph $D(G)$, where $D(G)$ specifies the dependence structure between the edges of G. The probability is given by:

$$P_G(X = x) = \frac{1}{c} \exp \left\{ \sum_{A \subseteq C(D)} \alpha(A) \right\},$$

where $\alpha(A)$ is an arbitrary constant if A is a clique of $D(G)$ and $\alpha(A) = 0$ otherwise, c is a normalizing constant, and $C(D)$ is the set of cliques of $D(G)$. Thus, the probability of any realization of G depends only on which cliques of $D(G)$ it contains.

The notion of a general Markov graph G was introduced by Frank and Strauss (1986) as a random graph whose dependence graph $D(G)$ contains no edges between disjoint sets of nodes, such as (i, j) and (k, l). This means that for a Markov graph G, edges that are not incident to the same node are conditionally independent. Thus, the cliques of the dependence graph of a Markov random graph correspond to sets of edges such that any pair of edges within

| Edge | Two-Star | Three-Star | Four-Star | Triangle |

Figure 7.25 Configurations for a Markov random graph.

the set must be incident to a node. Such sets are triangles and k-stars (or stars with k edges), $k = 1, 2, \ldots, N - 1$. A set or neighborhood of mutually conditionally dependent edges forms a configuration, and the configurations of an undirected Markov random graph are shown in Figure 7.25.

If we assume that the homogeneity condition holds (i.e., all isomorphic graphs have the same probability), then the PMF of a Markov random graph can be obtained in terms of the different configurations as follows:

$$p_G(x) = P_G(X = x) = \frac{1}{c}\exp\left\{ s_1(x)\theta + \sum_{k=2}^{N-1} s_k(x)\sigma_k + s_T(x)\tau \right\},$$

where:

- θ is the parameter of the edge statistic and $s_1(x)$ is the number of edges:

$$s_1(x) = \sum_{i=1}^{N}\sum_{j=i+1}^{N} x_{ij}$$

- σ_k is the parameter for the statistics of stars of size k (or k-stars), and $s_k(x)$ is the number of k-stars, $k \geq 2$, which is given by:

$$s_k(x) = \sum_{i=1}^{N}\binom{x_{i+}}{k},$$

where $\binom{x_{i+}}{k}$ is the number of k-stars in which node i is involved.

- τ is the parameter for the triangle statistic and $s_T(x)$ is the number triangles, which is:

$$s_T(x) = \sum_{i=1}^{N}\sum_{j=i+1}^{N}\sum_{h=j+1}^{N} x_{ij}x_{jh}x_{ih}.$$

7.16   MATRIX ALGEBRA OF GRAPHS

Several properties of a graph can be demonstrated via the matrix representation of the graph. One of the matrix representations of a graph is the *adjacency matrix* $A(x, y)$, which we have defined earlier in this chapter. Other matrix

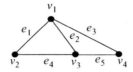

Figure 7.26   Example for matrix properties.

representations include *connection matrix, path matrix*, and *Laplacian matrix*. In this section we provide a brief discussion on these representations.

7.16.1   Adjacency Matrix

As defined earlier, the adjacency matrix $A(G)$ of a graph G is a matrix that has a value 1 in its (x, y) cell if vertices x and y are neighbors and zero otherwise, for all $x, y \in V$. Consider the graph in Figure 7.11a, which we reproduce in Figure 7.26.

The adjacency matrix is:

$$A = \begin{bmatrix} 0 & 1 & 1 & 1 \\ 1 & 0 & 1 & 0 \\ 1 & 1 & 0 & 1 \\ 1 & 0 & 1 & 0 \end{bmatrix}.$$

The matrix A^2 is given by:

$$A^2 = \begin{bmatrix} 0 & 1 & 1 & 1 \\ 1 & 0 & 1 & 0 \\ 1 & 1 & 0 & 1 \\ 1 & 0 & 1 & 0 \end{bmatrix} \begin{bmatrix} 0 & 1 & 1 & 1 \\ 1 & 0 & 1 & 0 \\ 1 & 1 & 0 & 1 \\ 1 & 0 & 1 & 0 \end{bmatrix} = \begin{bmatrix} 3 & 1 & 2 & 1 \\ 1 & 2 & 1 & 2 \\ 2 & 1 & 3 & 1 \\ 1 & 2 & 1 & 2 \end{bmatrix}.$$

The value of an off-diagonal entry in A^2, that is, ijth entry in A^2, $i \neq j$, is the number of vertices that are adjacent to both the ith and jth vertices, and is also the number of paths of length 2 between the ith and jth vertices. Similarly, the ith diagonal entry in A^2 is the number of 1's in the ith row (or column) of the matrix A. Thus, if the graph has no self-loops, the value of each diagonal entry of A^2 is equal to the degree of the corresponding vertex. This means that the trace of A^2 is twice the number of edges in the graph. Other topological properties of the adjacency matrix are as follows:

- In general, the ijth entry in A^k is the number of different edge sequences of k edges between vertex v_i and vertex v_j. That is, if we represent the ijth entry in A^k by A_{ij}^k, then A_{ij}^k is the number of walks of length k from vertex v_i to vertex v_j.
- A_{ii}^k is the number of closed walks of length k from vertex v_i to vertex v_i.

- The trace of A^k is the total number of closed walks of length k in the graph.
- The trace of A^3 is the six times the number of triangles in the graph.

7.16.2   Connection Matrix

The connection matrix $C(G)$ of a graph G with N vertices is the $N \times N$ matrix that has a value k in its (x, y) cell if the number of edges between vertices x and y is k for all $x, y \in V$. For an undirected graph, $C(G)$ is a symmetric matrix. For a directed graph, k is the number of tail-to-head edges between x and y, and the matrix is not symmetric. If there is no connection between vertices x and y, the entry in cell (x, y) is 0. For example, if G is the graph in Figure 7.26, then we have that:

$$C(G) = \begin{bmatrix} 0 & 1 & 1 & 1 \\ 1 & 0 & 1 & 0 \\ 1 & 1 & 0 & 1 \\ 1 & 0 & 1 & 0 \end{bmatrix}.$$

Note that for a simple graph, the connection matrix is identical to the adjacency matrix.

7.16.3   Path Matrix

A path matrix is defined for a pair of vertices in a graph. The rows of a path matrix $P(x, y)$ between vertex x and vertex y correspond to the different paths between x and y, and the columns correspond to the edges of the graph. The entry $p_{ij} = 1$ if the jth edge lies in the ith path between x and y, and $p_{ij} = 0$ otherwise. For example, consider all paths between v_2 and v_4 in Figure 7.26. There are three paths:

$$p_1 = \{e_1, e_3\},$$
$$p_2 = \{e_1, e_2, e_5\},$$
$$p_3 = \{e_4, e_5\}.$$

Thus, $P(v_2, v_4)$ is defined as follows:

$$P(v_2, v_4) = \begin{bmatrix} 1 & 0 & 1 & 0 & 0 \\ 1 & 1 & 0 & 0 & 1 \\ 0 & 0 & 0 & 1 & 1 \end{bmatrix}.$$

We can observe the following about $P(v_2, v_4)$:

a. The sum of each row is equal to the number of edges in the path it represents.
b. Each row must contain at least one unit entry.
c. A column with all unit entries represents an edge that belongs to every path between v_2 and v_4.
d. A column of all zeros represents an edge that does not lie on any path between v_2 and v_4.

7.16.4   Laplacian Matrix

The Laplacian matrix of graph G, $L(G)$, is a combination of the adjacency matrix and the degree matrix. Let D be a diagonal matrix whose entries d_{ii} are the degrees of the vertices. That is,

$$d_{ij} = \begin{cases} d(v_i) & i = j \\ 0 & \text{otherwise} \end{cases},$$

where $d(v_i)$ is the degree of vertex v_i. Then the Laplacian matrix is given by:

$$L = D - A.$$

For example, for the graph in Figure 7.26, we have that:

$$A = \begin{bmatrix} 0 & 1 & 1 & 1 \\ 1 & 0 & 1 & 0 \\ 1 & 1 & 0 & 1 \\ 1 & 0 & 1 & 0 \end{bmatrix}, \quad D = \begin{bmatrix} 3 & 0 & 0 & 0 \\ 0 & 2 & 0 & 0 \\ 0 & 0 & 3 & 0 \\ 0 & 0 & 0 & 2 \end{bmatrix}, \quad L = D - A = \begin{bmatrix} 3 & -1 & -1 & -1 \\ -1 & 2 & -1 & 0 \\ -1 & -1 & 3 & -1 \\ -1 & 0 & -1 & 2 \end{bmatrix}.$$

The Laplacian matrix can be used to find other properties of the graph, such as the spectral gap that is discussed later in this chapter.

7.17   SPECTRAL PROPERTIES OF GRAPHS

The spectrum of a graph G with N vertices is the set of eigenvalues $\lambda_i, i = 1, \ldots, N$, of its adjacency matrix A. The spectral density of the graph is defined by:

$$\rho(\lambda) = \frac{1}{N} \sum_{i=1}^{N} \delta(\lambda - \lambda_i),$$

where $\delta(x)$ is the Dirac delta function. The number λ is an eigenvalue of matrix A if there is a nonzero vector x satisfying:

$$Ax = \lambda x.$$

Each such vector x is called the *eigenvector* of A corresponding to the eigenvalue λ. The eigenvalues of the adjacency matrix A are often referred to as the eigenvalues of G, and the collection of eigenvalues of G is referred to as the *spectrum* of G. Note that eigenvalues are the solutions of $\det(A - \lambda I) = 0$, where I is the identity matrix. Alternatively, the eigenvalues are the values of λ that make the matrix $A - \lambda I$ singular.

The spectral density approaches a continuous function as $N \to \infty$. When G is undirected, without loops or multiple edges, A is real and symmetric. This means that G has real eigenvalues $\lambda_1 \leq \lambda_2 \leq \ldots \leq \lambda_N$, and the eigenvectors corresponding to distinct eigenvalues are orthogonal. Equality between eigenvalues corresponds to eigenvalue multiplicity. When G is directed, the eigenvalues can have imaginary part, and the ordering and properties of eigenvalues and eigenvectors become more complicated.

The topological properties of graphs, such as their patterns of connectivity, can be analyzed using spectral graph theory. For example, the kth moment of the spectral density is given by:

$$M_k = \frac{1}{N} \sum_i (\lambda_i)^k.$$

The quantity $D_k = NM_k$ is the total number of walks of length k from each node of the graph to itself. The eigenvalues also provide a series of bounds on the diameter of a graph. For example, the diameter of G is less than the number of distinct eigenvalues of G. Also, for a graph with all degrees equal to k, the second largest eigenvalue, λ_{N-1}, can be used to obtain an upper bound of the diameter. Furthermore, a graph G with at least one edge is a bipartite graph if its spectrum is symmetric with respect to 0; that is, if λ is an eigenvalue of G, then $-\lambda$ is also an eigenvalue of G. Alternatively, in a bipartite graph, all nonzero eigenvalues come in pairs with their negatives. Other properties of eigenvalues include:

- The sum of all eigenvalues, including multiplicities, is equal to the trace of the adjacency matrix A; that is,

$$\sum_{i=1}^{N} \lambda_i = \sum_{i=1}^{N} a_{ii} = Tr(A).$$

- The sum of all eigenvalues of an adjacency matrix is 0, which follows from the fact that this sum is the trace of A that from the previous bullet is 0 since $a_{ii} = 0$ for all i.
- The product of all eigenvalues, including multiplicities, is equal to the determinant of the adjacency matrix A; that is,

$$\prod_{i=1}^{N} \lambda_i = \det(A).$$

The number of nonzero eigenvalues, including multiplicities, is the rank of A. One implication of this property is the following. The complete bipartite graph K_{mn} has an adjacency matrix of rank 2. Thus, we expect to have eigenvalue 0 of multiplicity $N - 2$, and two nontrivial eigenvalues. These nontrivial eigenvalues should be equal to $\pm\lambda$ since the sum of all eigenvalues is always 0.

7.17.1   Spectral Radius

The *spectral radius* of G is the largest eigenvalue of G that is distinct from 1. For example, the adjacency matrix of the graph in Figure 7.26 is:

$$A = \begin{bmatrix} 0 & 1 & 1 & 1 \\ 1 & 0 & 1 & 0 \\ 1 & 1 & 0 & 1 \\ 1 & 0 & 1 & 0 \end{bmatrix}.$$

To obtain the eigenvalues we proceed as follows. First, we solve the equation:

$$\det(A - \lambda I) = \begin{vmatrix} -\lambda & 1 & 1 & 1 \\ 1 & -\lambda & 1 & 0 \\ 1 & 1 & -\lambda & 1 \\ 1 & 0 & 1 & -\lambda \end{vmatrix} = \lambda^4 - 5\lambda^2 - 4\lambda = \lambda(\lambda+1)(\lambda^2 - \lambda - 4) = 0.$$

The roots of this *characteristic polynomial* are $\{-1.56155, -1, 0, 2.56155\}$. Since the largest nontrivial root is 2.56155, the spectral density is 2.56155.

7.17.2   Spectral Gap

The spectral gap of a graph is the largest nontrivial eigenvalue of its Laplacian matrix. The eigenvalues of the Laplacian matrix of the graph in Figure 7.26 are obtained from:

$$\det(L - \lambda I) = \begin{vmatrix} 3-\lambda & -1 & -1 & -1 \\ -1 & 2-\lambda & -1 & 0 \\ -1 & -1 & 3-\lambda & -1 \\ -1 & 0 & -1 & 2-\lambda \end{vmatrix} = \lambda^4 - 10\lambda^3 + 32\lambda^2 - 32\lambda$$

$$= \lambda(\lambda-2)(\lambda-4)^2 = 0.$$

Thus, the eigenvalues of the Laplacian matrix are $\{0, 2, 4, 4\}$. Since the largest eigenvalue is 4, the spectral gap is 4. In general, because L is symmetric, its eigenvalues are real and nonnegative and are such that $0 = \lambda_1 \leq \lambda_2 \leq \ldots \leq \lambda_n$.

Also, the trace of L, $Tr(L)$, is the sum of the eigenvalues, which is twice the number of edges in G.

7.18 GRAPH ENTROPY

Entropy is a measure of the uncertainty associated with a random variable. It is popularly used in the field of information theory. If a random variable has a uniform distribution, the states of system are highly disordered and the entropy of the probability distribution increases. On the contrary, when the probability distribution of a random variable is not a uniform distribution, some of the states can be predictable and the states of the system are more ordered, which results in lower entropy. Thus, entropy is a helpful tool that can used to describe the state of order in a system.

The entropy of a graph is a measure of the structure of the graph and may be interpreted as the degree of randomness of the graph. The more random a graph, the higher is its entropy. Consider the graph $G = (V, E)$ with a probability distribution P on its vertex set of $V(G)$, where $p_i \in [0, 1]$ is the probability of vertex $v_i, i = 1, \ldots, N$. The entropy of the graph G is given by:

$$H(G, P) = -\sum_{i=1}^{N} p_i \log_2 (p_i).$$

$p_i = p(v_i)$ can be defined in any desired manner. Thus, we write:

$$p_i = p(v_i) = \frac{f(v_i)}{\sum\limits_{i=1}^{N} f(v_i)},$$

where $f(v_i)$ is an arbitrary function of v_i. For example, we may define $f(v_i) = d(v_i)$, where $d(v_i)$ is the degree of vertex v_i.

7.19 DIRECTED ACYCLIC GRAPHS

Directed acyclic graphs (DAGs) are directed graphs that have no cycles. If an edge $e_{ij} = (v_i, v_j)$ exists from node v_i to node v_j, then v_i is called the *parent* of v_j and v_j is called the *child* of v_i. That is, the edge points from the parent to the child. If v_i is the parent of v_j and v_j is the parent of v_k, then we say that v_i is an *ancestor* of v_k and v_k is a *descendant* of v_i. An example of a DAG is shown in Figure 7.27.

In Figure 7.27, node v_0 is the parent of node v_1, which in turn is the parent of node v_4. Thus v_0 is an ancestor of v_4 and v_4 is a descendant of v_0. Also, since node v_4 and node v_5 are children of node v_2, they are *siblings*. DAGs are widely used in Bayesian networks.

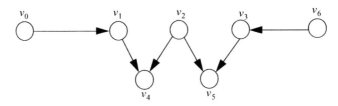

Figure 7.27   A DAG.

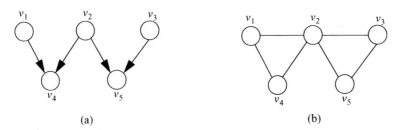

Figure 7.28   (a) Directed graph and (b) corresponding moral graph.

7.20   MORAL GRAPHS

A moral graph is used to convert a problem that is specified using a directed graph to one that is specified using an undirected graph. It is derived from a directed graph by adding extra edges between all unconnected pairs of nodes that have a common child and dropping the arrows. Thus, the goal is to "marry" the parents of each node. An example is shown in Figure 7.28 where Figure 7.28a is the original directed graph and Figure 7.28b is the corresponding moral graph.

In Figure 7.28a, nodes v_1 and v_2 are both parents of node v_4; similarly, nodes v_2 and v_3 are the parents of node v_5. Thus, an edge is drawn to connect node v_1 and v_2, and an edge is drawn to connect nodes v_2 and v_3. Finally all the arrows are dropped to produce Figure 7.28b. The process of adding an extra edge between two parents of a child is called *moralization*.

7.21   TRIANGULATED GRAPHS

An undirected graph is called a triangulated graph if every cycle of length greater than 3 has an edge joining two nonadjacent vertices of the cycle. The edge joining the two vertices is called a *chord*, and triangulated graphs are also called *chordal graphs*. For example, Figure 7.29 is a triangulated graph.

Let $\aleph(v)$ denote the set of neighbors of node v. Node v is called a *simplicial node* if $\aleph(v)$ is a clique. Let G be a triangulated graph and let v be a simplicial

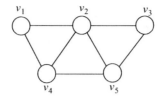

Figure 7.29   A triangulated graph.

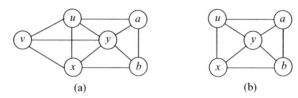

(a)   (b)

Figure 7.30   A triangulated graph with (a) simplicial node v; (b) with simplicial node removed.

node in G. Let G_S be the graph resulting from eliminating v from G. Then G_S is a triangulated graph. An example is illustrated in Figure 7.30 where $\aleph(v) = \{u, x, y\}$.

Let C denote the set of cliques from an undirected graph G, and let the cliques of C be organized in a tree T. If A and B are a pair of nodes in T and all nodes on the path between A and B contain the intersection $A \cap B$, then T is called a *join tree*. It is well known that if the cliques of an undirected graph can be organized into a join tree, then the graph is a triangulated graph.

7.22   CHAIN GRAPHS

Directed graphs are used to model causal relationships whereby the direction of an edge indicates the direction of influence or causal direction. Sometimes the variables in a model can be grouped into blocks such that the variables in a block are not ordered but there is an ordering between blocks. An acyclic graph that contains both directed and undirected edges is called a chain graph. In this case all cycles in the graph consist of only undirected edges. An example of a chain graph is given in Figure 7.31 where the graph consists of two blocks

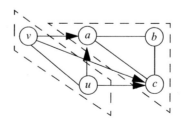

Figure 7.31   Example of a chain graph.

of nodes: $\{a, b, c\}$ and $\{u, v\}$. The edges within dotted areas are undirected while edges between dotted areas are directed edges.

7.23 FACTOR GRAPHS

A factor graph represents factorizations of arbitrary multivariate functions and dependencies among variables. Consider the function $f(x_1, \ldots, x_n)$. Let it factor into a product of several *local functions* each of which has a subset of $\{x_1, \ldots, x_n\}$ as its arguments; that is,

$$f(x_1, \ldots, x_n) = \prod_{k \in K} f_k(X_k),$$

where K is a discrete index set, X_k is a subset of $\{x_1, \ldots, x_n\}$, and $f_k(X_k)$ is a local function (or a factor) having X_k as arguments.

A factor graph is a bipartite graph that expresses the structure of the preceding factorization. The two sets of nodes are the *variable nodes* that consist of one node for each variable x_i and the *factor nodes* that consist of one node for each factor f_k. An edge connects variable node x_i to factor node f_k if and only if x_i is an argument of f_k. For example, assume that:

$$f(x_1, x_2, x_3, x_4, x_5, x_6) = f_A(x_1, x_3) f_B(x_2, x_4) f_C(x_3, x_4, x_5) f_D(x_5, x_6).$$

That is, $K = \{A, B, C, D\}$, $X_A = \{x_1, x_3\}$, $X_B = \{x_2, x_4\}$, $X_C = \{x_3, x_4, x_5\}$, and $X_D = \{x_5, x_6\}$. Then the factor graph is shown in Figure 7.32 where the variable nodes are represented by circles and the factor nodes are represented by rectangles.

An alternative method of representing a factor graph is the so-called Forney-style factor graph (FFG), which was introduced by Forney (2001) as a "normal graph." FFG consists of the following:

- *Nodes:* There is a node for every factor f_k.
- *Edges:* There is an edge or a half-edge for every variable. Half-edges connect to one node only and an edge is connected to two nodes.

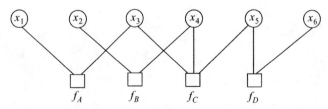

Figure 7.32   Example of a factor graph.

- *Connections:* An edge (or half-edge) is connected to a node if and only if the factor (corresponding to the node) is a function of the variable (corresponding to the edge).
- *Number of Connections per Edge:* An edge is connected to no more than two nodes.
- *Configuration:* A configuration is a particular assignment of values to all variables.

The FFG for the function $f(x_1, x_2, x_3, x_4, x_5, x_6) = f_A(x_1, x_3)f_B(x_2, x_4)f_C(x_3, x_4, x_5)f_D(x_5, x_6)$ is shown in Figure 7.33.

Factor graphs are used in the sum-product algorithm that is discussed in Chapter 8. They are a unifying framework for a wide variety of system models in coding, signal processing, and artificial intelligence.

Consider the random variables $X_1, X_2, \ldots, X_5$ whose joint PMF can be written as follows:

$$P[X_1, X_2, X_3, X_4, X_5] = P[X_1]P[X_2|X_1, X_5]P[X_3|X_1, X_2]P[X_4|X_3]P[X_5].$$

The joint PMF can be represented by the DAG shown in Figure 7.34.

The DAG can be converted into a factor graph as shown in Figure 7.35 where the unconditional and conditional probabilities constitute the factor

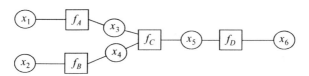

Figure 7.33 Example of an FFG.

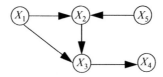

Figure 7.34 DAG example.

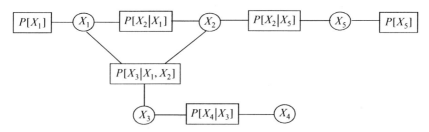

Figure 7.35 An FFG for the DAG of Figure 7.34.

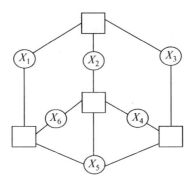

Figure 7.36   An FFG for the system of equations.

nodes. This association between DAGs and factor graphs will be discussed in greater detail in Chapter 8.

   Another example of the application of factor graphs is in error-correcting codes. Consider the following system of equations (see Wiberg 1996):

$$x_1 + x_2 + x_3 = 0,$$
$$x_3 + x_4 + x_5 = 0,$$
$$x_1 + x_5 + x_6 = 0,$$
$$x_2 + x_4 + x_6 = 0.$$

The variables are binary variables that take the values of 0 or 1, and addition is done modulo 2. This system of equations can be expressed as the factor graph shown in Figure 7.36 where a square together with its neighbors corresponds to an equation with its variables.

7.24   PROBLEMS

7.1   Consider an assignment problem represented by the following matrix:

$$A = \begin{bmatrix} 22 & 15 & 23 & 27 \\ 30 & 15 & 17 & 28 \\ 25 & 20 & 25 & 32 \end{bmatrix}.$$

   a. Obtain the minimum cost assignment using the Hungarian method.

   b. Obtain the maximum profit assignment using the Hungarian method.

7.2   Consider the weighted graph shown in Figure 7.37. Use the Kruskal's algorithm to obtain the minimum weight spanning tree.

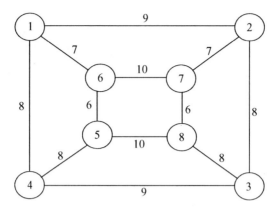

Figure 7.37 Graph for Problem 7.2.

7.3 Five women labeled A, B, C, D, and E, and six men numbered 1, 2, 3, 4, 5, and 6 are at a party. The men that each woman would like to dance with are displayed by means of the bipartite graph shown in Figure 7.38. Is it possible to have all women dancing at the same time so that each woman is dancing with a man that she likes to dance with?

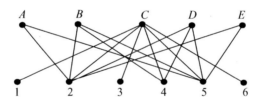

Figure 7.38 Figure for Problem 7.3.

7.4 Calculate the spectral radius and the spectral gap of the graph shown in Figure 7.39.

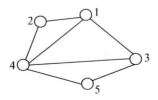

Figure 7.39 Figure for Problem 7.4.

7.5 Consider the DAG shown in Figure 7.40 and assume that the joint PMF of the nodes is given by:

$$P[v_0, v_1, v_2, v_3, v_4, v_5, v_6] = P[v_0]P[v_2]P[v_6]P[v_1|v_0]P[v_3|v_6]$$
$$P[v_4|v_1, v_2]P[v_5|v_2, v_3].$$

Obtain the factor graph of the DAG.

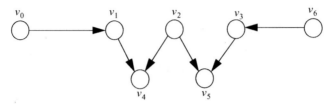

Figure 7.40 Figure for Problem 7.5.

7.6 Consider the DAG shown in Figure 7.41.
   a. Obtain the moral graph of the DAG.
   b. Obtain the cliques of the moral graph.

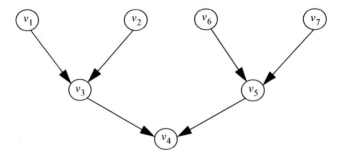

Figure 7.41 Figure for Problem 7.6.

8

BAYESIAN NETWORKS

8.1 INTRODUCTION

A probabilistic network, which is sometimes referred to as a probabilistic graphical model, is a tool that enables us to visually illustrate and work with conditional independencies among variables in a given problem. Specifically, nodes represent variables and the lack of an edge between two nodes represents conditional independence between the variables. There are two types of graphical models: *directed graphical models* and *undirected graphical models*. Undirected graphical models are called *Markov random fields* or *Markov networks* and are popularly used in the physics and vision communities. Directed graphical models have no directed cycles and are called *Bayesian networks* (BNs) or *belief networks*; they are popularly used in the artificial intelligence and statistics communities.

In an undirected graphical model, two nodes A and B are defined to be conditionally independent given a third node C, written $A \perp B|C$, if all paths between A and B are separated by C. If the joint distribution of A, B, and C is known, then we may write:

$$P[A, B|C] = P[A|B, C]P[B|C] = P[A|C]P[B|C].$$

In a directed graphical model, conditional independence can be displayed graphically. For example, consider the distribution:

Fundamentals of Stochastic Networks, First Edition. Oliver C. Ibe.
© 2011 John Wiley & Sons, Inc. Published 2011 by John Wiley & Sons, Inc.

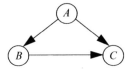

Figure 8.1   Directed graphical model of the probability distribution.

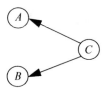

Figure 8.2   Directed graphical model of conditional independence.

$$P[A, B, C] = P[C|A, B]P[B|A]P[A].$$

For each conditional distribution we add a directed arc from the node corresponding to the variable on which the distribution is conditioned. Thus, for $P[C|A, B]$ we have an arc from A to C and another arc from B to C. Since $P[A]$ is an unconditional probability, there is no arc coming into node A. Thus, the directed graphical model for the joint distribution is as shown in Figure 8.1.

The directed graphical model for the conditional independence $A \perp B|C$ is given in Figure 8.2. This means that once the state of C is known, no knowledge of A will alter the probability of B.

In this chapter we discuss only the directed graphical networks, which are BNs.

8.2   BAYESIAN NETWORKS

BNs are a combination of graph theory and probability theory. They encode one's beliefs for a system of variables and are used in various domains including automatic speech recognition. Recall that the Bayes' theorem states that:

$$P[X|Y] = \frac{P[X, Y]}{P[Y]} = \frac{P[Y|X]P[X]}{P[Y]}.$$

This means that if we know the *prior probability* $P[X]$, which represents our belief about X, and later observe Y, then our revised belief about X, which is the *posterior probability* $P[X|Y]$, is obtained by multiplying the prior probability $P[X]$ by the factor $P[Y|X]/P[Y]$. The quantity $P[Y|X]$ is called the *likelihood function*. Since $P[Y]$ is a known quantity and thus is constant, we may write:

Figure 8.3 Graphical model of Bayes' theorem.

$$P[X|Y] \propto P[Y|X] \times P[X]$$
$$\Rightarrow \text{Posterior Probability} \propto \text{Likelihood Function} \times \text{Prior Probability,}$$
$$P[X|Y] = \eta P[Y|X] \times P[X],$$

where $\eta = 1/P[Y]$ is a constant of proportionality. This relationship between the prior probability and posterior probability can be displayed graphically as shown in Figure 8.3 that shows a directed arc from X to Y. This indicates that the joint probability $P[X, Y]$ is decomposed into the prior probability and $P[Y|X]$. Thus, we may visualize this as follows: X is a possible cause that has an effect on Y.

A BN is a directed acyclic graph in which each node represents a random variable and each set of arcs into a node represents a probabilistic dependency between the node and its parents (i.e., the nodes at the other ends of the incoming arcs). Thus, the network represents the conditional independence relations among the variables in the network, and the absence of an arc between two nodes implies a conditional independence between the two nodes. Consider a probability distribution $P[X]$ over a set of random variables $X = \{X_1, \ldots, X_n\}$. We can repetitively decompose the distribution into a product of conditional probability distributions (CPDs) as follows:

$$P[X] = P[X_1, X_2, \ldots, X_n] = P[X_1|X_2, \ldots, X_n]P[X_2, \ldots, X_n]$$
$$= P[X_1|X_2, \ldots, X_n]P[X_2|X_3, \ldots, X_n]\ldots P[X_{n-1}|X_n]P[X_n]$$
$$= \prod_{i=1}^{n} P[X_i|X_{i+1}, \ldots, X_n].$$

For example, let $n = 4$, and we obtain:

$$P[X] = P[X_1|X_2, X_3, X_4]P[X_2|X_3, X_4]P[X_3|X_4]P[X_4].$$

The directed acyclic graph corresponding to this equation is shown in Figure 8.4.

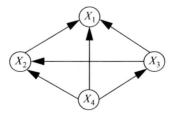

Figure 8.4 DAG of $P[X]$.

In general, a BN is specified as the pair (G, Θ), where G is the directed acyclic graph (DAG) whose vertices correspond to the random variables $X_1; \ldots, X_n$ and Θ is the CPD of these random variables. Each parameter θ_i of Θ is specified by $\theta_i = P[X_i|\text{Pa}(X_i)]$, where $\text{Pa}(X_i)$ is the set of parents of X_i. Thus, a BN specifies a unique joint probability distribution over all the random variables as follows:

$$P[X] = P[X_1, X_2, \ldots, X_n] = \prod_{i=1}^{n} P[X_i|\text{Pa}(X_i)] = \prod_{i=1}^{n} \theta_i.$$

This means that the joint probability of all the variables is the product of the probabilities of each variable given its parents' values. Thus, in a BN, directed edges indicate which other variables a given variable is conditioned upon. For this reason the edges in BNs can be considered causal connections where each parent node causes an effect on its children. Note that if X_i has no parents, then we obtain the unconditional probability $P[X_i|\text{Pa}(X_i)] = P[X_i]$. Also, if $X_j \in \text{Pa}(X_i)$, we say that $X_i \in \text{Ch}(X_j)$, where $\text{Ch}(X_j)$ is the set of children of X_j. Another relationship associated with a variable X_i is the *Markov blanket*, which consists of all the parents of X_i, all the children of X_i, and all its children's other parents, called *co-parents* of its children. The Markov blanket of X_i, $MB(X_i)$, contains all the variables that shield X_i from the rest of the network. This is illustrated in Figure 8.5.

The Markov blanket of a node is the only knowledge needed to predict the behavior of that node. Every set of nodes in the network is conditionally independent of X_i when conditioned on $MB(X_i)$; that is,

$$P[X_i|MB(X_i), Y] = P[X_i|MB(X_i)].$$

A BN is used to represent knowledge about an uncertain domain. Thus, given a BN for a domain and evidence about a particular situation in that domain, conclusions can be drawn about that situation. Evidence is an assignment of values to a set of the random variables. Once new evidence is obtained the BN computes new probabilities based on the new information.

Consider the following BN that illustrates a situation where there are multiple causes for one effect. It is known that the common cold (C) causes fever

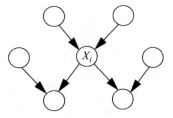

Figure 8.5 Markov blanket of Node X_i.

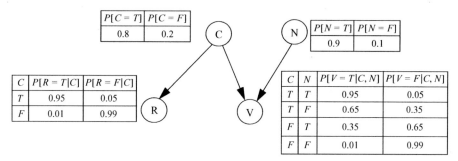

Figure 8.6 Example of a BN.

(V) and runny nose (R), and pneumonia (N) causes fever. Figure 8.6 shows the DAG as well as the conditional distributions for the problem. The entries T and F denote "true" and "false," respectively. The conditional probabilities associated with each random variable are usually stored in tables called *conditional probability tables* (CPTs).

The concept of conditional independence helps to simplify computations in the BNs. For example, suppose a BN consists of the variable or node set $X = \{X_1, \ldots, X_n\}$, and suppose the variables $Y_1, \ldots, Y_k$ are observed to have the values $y_1, \ldots, y_k$, respectively. Assume that the conditional probability of X is queried given the observations. We have that:

$$P[X|Y_1 = y_1, \ldots, Y_k = y_k] = \frac{P[X, Y_1 = y_1, \ldots, Y_k = y_k]}{P[Y_1 = y_1, \ldots, Y_k = y_k]}.$$

Now, the denominator is given by:

$$P[Y_1 = y_1, \ldots, Y_k = y_k] = \sum_{v \in domain(X)} P[X = v, Y_1 = y_1, \ldots, Y_k = y_k].$$

Using the product rule of probability, both the numerator and denominator can be factored as follows:

$$P[X|Y_1 = y_1, \ldots, Y_k = y_k] = \frac{P[X, Y_1 = y_1, \ldots, Y_k = y_k]}{\sum_{v \in domain(X)} P[X = v, Y_1 = y_1, \ldots, Y_k = y_k]}$$

$$= \frac{P[Y_1|Y_2, \ldots, Y_k, X]P[Y_2|Y_3, \ldots, Y_k, X] \ldots P[Y_k|X]P[X]}{\sum_{v \in domain(X)} P[Y_1|Y_2, \ldots, Y_k, X]P[Y_2|Y_3, \ldots, Y_k, X] \ldots P[Y_k|X]P[X]}.$$

Due to conditional independence, we can remove certain variables from the conditional list by recognizing that if z_0 is not a child of z_n in the BN, then:

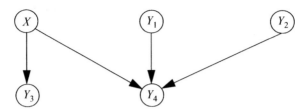

Figure 8.7 Example of a BN.

$$P[z_0|z_1,\ldots,z_{n-1},z_n,z_{n+1},\ldots z_k] = P[z_0|z_1,\ldots,z_{n-1},z_{n+1},\ldots z_k].$$

For example, consider the BN shown in Figure 8.7. In this case, we have that:

$$
\begin{aligned}
P[X|Y_1,Y_2,Y_3,Y_4] &= \frac{P[Y_1|Y_2,Y_3,Y_4,X]P[Y_2|Y_3,Y_4,X]P[Y_4|X]P[X]}{\displaystyle\sum_{v\in domain(X)} P[Y_1|Y_2,Y_3,Y_4,X]P[Y_2|Y_3,Y_4,X]P[Y_4|X]P[X]} \\[2mm]
&= \frac{P[Y_1]P[Y_2]P[Y_3|X]P[Y_4|X]P[X]}{\displaystyle\sum_{v\in domain(X)} P[Y_1]P[Y_2]P[Y_3|X]P[Y_4|X]P[X]} \\[2mm]
&= \frac{P[Y_3|X]P[Y_4|X]P[X]}{\displaystyle\sum_{v\in domain(X)} P[Y_3|X]P[Y_4|X]P[X]}.
\end{aligned}
$$

8.3 CLASSIFICATION OF BNs

There are several ways to classify BNs. From the perspective of network structure, BNs can be either *singly connected* or *multiply connected*. Singly connected BNs are also called *polytrees* and are characterized by the fact that there is at most one path between any two nodes. In a multiply connected BN, there is at least one pair of nodes that has more than one path between them. Examples of singly connected and multiply connected BNs are shown in Figure 8.8.

BNs can also be classified according to the type of random variable in the model. Thus, they may be classified as discrete BNs, continuous BNs, and hybrid BNs, which contain both discrete and continuous random variables.

Also, from time point of view, BNs may be classified as static BNs, in which nodes do not change their values over time, and as dynamic BNs, where the nodes are allowed to change their values over time. It is also possible to have partially dynamic BNs where some of the nodes do not change their values over time and others change their values over time.

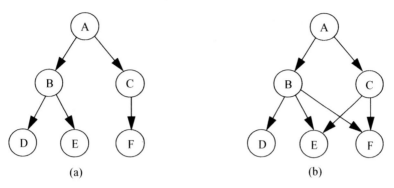

Figure 8.8   Examples of (a) singly connected and (b) multiply connected BNs.

8.4   GENERAL CONDITIONAL INDEPENDENCE AND *d*-SEPARATION

The graphical property of *d*-separation can be used to determine conditional independence in a BN. The "d" in *d*-separation stands for dependence. Specifically, if two sets of nodes X and Y are *d*-separated in a graph by a third set Z, then the corresponding variable sets X and Y are conditionally independent given the variables in Z. Alternatively, two sets of nodes X and Y are *d*-separated in a BN by a third set Z if and only if every path p between X and Y is "blocked" by a member of Z, where the term "blocked" means that there is an intermediate variable $v \in Z$ such that:

1. p contains a chain $x \rightarrow v \rightarrow y$ such that the middle node $v \in Z$, where $x \in X$ and $y \in Y$; this is called a tail-to-head or serial pattern
2. p contains a chain $x \leftarrow v \leftarrow y$ such that the middle node $v \in Z$, where $x \in X$ and $y \in Y$; this is called a head-to-tail pattern
3. p contains a fork $x \leftarrow v \rightarrow y$ such that the middle node $v \in Z$, where $x \in X$ and $y \in Y$; this is called a tail-to-tail or diverging pattern
4. p contains an inverted fork $x \rightarrow v \leftarrow y$ such that the middle node $v \notin Z$ and no descendant of v is in Z, where $x \in X$ and $y \in Y$. This is called a head-to-head pattern.

Thus, in both 1 and 2 above, $v \in Z$ and has one arrow on the path leading in and one arrow leading out. In 3, $v \in Z$ and has both path arrows leading out. In 4, neither v nor any of its descendants is in Z and both path arrows lead into v. These observations are summarized in Figure 8.9.

8.5   PROBABILISTIC INFERENCE IN BNs

One of the purposes of constructing BNs is to perform a probabilistic inference. Inference refers to the process of computing the posterior probability

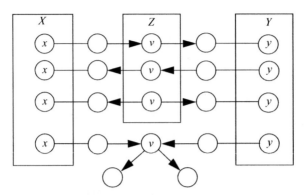

Figure 8.9   *Z d*-separates *X* and *Y*.

distribution $P[X|Y = y]$ of a set X of query variables after obtaining some observation $Y = y$, where Y is a list of observed variables. The most common questions that can be asked in a BN are the marginal probability of a variable X, the conditional probability of X given Y, and the joint probability of a set of variables in the network.

There are two types of inference algorithms: exact algorithms and approximate algorithms. Exact algorithms work by storing locally on each node a belief distribution for the possible values, and by propagating messages among adjacent nodes of the network. These messages are used to update the belief distribution of a node that receives them. When the belief state of a node changes, it generates a message to its neighbors, and the process is repeated until convergence is reached for the values of the belief distributions. Examples of exact algorithms include:

a. Variable elimination
b. Belief propagation, which is a special case of the sum-product algorithm
c. Clique tree propagation, which is also called the junction tree algorithm

Approximate algorithms use the BN as an instance generator to produce an adequate sample from which the result is estimated as the relative frequency of instances that fulfill specific conditions. Examples of approximate algorithms include:

a. Stochastic sampling and Markov chain Monte Carlo
b. Variational methods
c. Loopy belief propagation

However, it must be emphasized that both exact inference and approximate inference for BNs are NP-hard, which means that they are harder to solve than problems that can be solved by nondeterministic Turing machines in polynomial time. In the remainder of this section we consider the sum-product algorithm and the junction tree algorithm, both of which use the graph theoretic concepts that are presented in Chapter 7.

8.5.1 The Sum-Product Algorithm

The sum-product algorithm performs inference by passing messages along the edges of the graph. The message arriving at a node (or random variable) is a probability distribution (or a function that is proportional to the probability distribution) that represents the inference for the node as given by the part of the graph that the message came from. The algorithm is known to be exact if the graph is a tree.

Suppose we want to obtain the marginal probability distribution $P[X_i]$ from the joint probability distribution $P[X] = P[X_1, \ldots, X_{i-1}, X_i, X_{i+1}, \ldots X_n]$. This can be obtained by summing $P[X]$ over all variables except X_i; that is,

$$P[X_i] = \sum_{X \setminus X_i} P[X] = \sum_{X \setminus X_i} \prod_{i=1}^{n} P[X_i | \mathrm{Pa}(X_i)] = \sum_{X \setminus X_i} \prod_{i=1}^{n} f_i(X_i),$$

where $X \setminus X_i$ denotes the set of variables in X with X_i omitted and $f_i(X_i) = P[X_i | \mathrm{Pa}(X_i)]$. Thus, we can substitute for $P[X]$ using a factor graph. Recall from Chapter 7 that a factor graph is a bipartite graph that expresses the function $f(x_1, \ldots, x_n)$ as a product of several local functions (or factors), each of which has a subset of $\{x_1, \ldots, x_n\}$ as arguments; that is,

$$f(x_1, \ldots, x_n) = \prod_{k \in K} f_k(X_k),$$

where K is a discrete index set, X_k is a subset of $\{x_1, \ldots, x_n\}$, and $f_k(X_k)$ is a local function having X_k as arguments.

Interchanging the sum and product in the computation of $P[X_i]$ we have that:

$$P[X_i] = \prod_{i=1}^{n} \left\{ \sum_{X \setminus X_i} f_i(X_i) \right\}.$$

The sum-product algorithm can compute the marginals of all variables (or nodes) efficiently and exactly if the factor graph is a tree, which implies that there is only one path between any two nodes. It involves passing messages on the factor graph. There are two types of messages passed, which are:

1. A message from a factor node f_s to a variable node x, which is denoted by $m_{f_s \to x}(x)$.
2. A message from a variable node x_k to a factor node f_s, which is denoted by $m_{x_k \to f_s}(x_k)$.

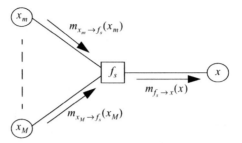

Figure 8.10 Factorization of subgraph associated with factor node f_s.

A message sent from a node u on edge e is the product of the factor at u (or one unit, if u is a variable node) and all messages received at u on edges other than e. Thus the messages are defined recursively as follows:

$$m_{f_s \to x}(x) = \sum_{x_1} \cdots \sum_{x_M} f_s(x, x_1, \dots, x_M) \prod_{m \in \aleph(f_s) \backslash x} m_{x_m \to f_s}(x_m),$$

$$m_{x_m \to f_s}(x_m) = \prod_{l \in \aleph(x_m) \backslash f_s} m_{f_l \to x_m}(x_m),$$

where $\aleph(f_s)$ is the set of variable nodes that are neighbors of the factor node f_s, $\aleph(f_s) \backslash x$ is the same set but with variable node x removed, $\aleph(x_m)$ is the set of neighbors of variable node x_m, and $\aleph(x_m) \backslash f_s$ is the same set but with factor node f_s removed. Figure 8.10 illustrates the factorization of the subgraph associated with the factor node f_s.

The goal is to calculate the marginal distribution of variable node x, and this distribution is given by the product of incoming messages along all the edges that are incident on the node. The recursion is initialized by picking an arbitrary node as the root of the tree, and we start all the messages at the leaf nodes. If a leaf node is a variable node, then the message that it sends on the only edge to a factor node f is given by:

$$m_{x \to f}(x) = 1.$$

If the leaf node is a factor node, it sends to a variable node the message:

$$m_{f \to x}(x) = f(x).$$

A node can send out a message if all the necessary incoming messages have been received. After all the messages have been sent out, the desired marginal probabilities can be computed as:

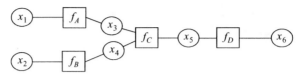

Figure 8.11 Example of the algorithm.

$$P[x] = \prod_{f \in \aleph(x)} m_{f \to x}(x).$$

The marginal distributions of the set of variables x_s involved in factor f_s can also be computed as the product of messages arriving at the factor and the local factor at the node as follows:

$$P[x] = f_s(x_s) \prod_{x \in \aleph(f_s)} m_{x \to f_s}(x).$$

As an example of the use of the algorithm, consider the factor graph shown in Figure 8.11.

Assume that x_6 is the root, which means that x_1 and x_2 are leaf nodes. Thus, the sequence of messages is as follows:

$$m_{x_1 \to f_A}(x_1) = 1,$$

$$m_{f_A \to x_3}(x_3) = \sum_{x_1} f_A(x_1, x_3),$$

$$m_{x_2 \to f_B}(x_2) = 1,$$

$$m_{f_B \to x_4}(x_4) = \sum_{x_2} f_B(x_2, x_4),$$

$$m_{x_3 \to f_C}(x_3) = m_{f_A \to x_3}(x_3),$$

$$m_{x_4 \to f_C}(x_4) = m_{f_B \to x_4}(x_4),$$

$$m_{f_C \to x_5}(x_5) = \sum_{x_3} \sum_{x_4} f_C(x_3, x_4, x_5) \prod_{m \in \aleph(f_C) \backslash x_5} m_{x_m \to f_C}(x_m)$$

$$= \sum_{x_3} \sum_{x_4} f_C(x_3, x_4, x_5) m_{x_3 \to f_C}(x_3) m_{x_4 \to f_C}(x_4),$$

$$m_{x_5 \to f_D}(x_5) = m_{f_D \to x_6}(x_6),$$

$$m_{f_D \to x_6}(x_6) = \sum_{x_5} f_D(x_5, x_6) m_{x_5 \to f_D}(x_5).$$

Next, we consider messages that are propagated from the root to the leaf nodes, which are as follows:

$$m_{x_6 \to f_D}(x_6) = 1,$$

$$m_{f_D \to x_5}(x_5) = \sum_{x_6} f_D(x_5, x_6),$$

$$m_{x_5 \to f_C}(x_5) = m_{f_D \to x_5}(x_5),$$

$$m_{f_C \to x_4}(x_4) = \sum_{x_3} \sum_{x_5} f_C(x_3, x_4, x_5) m_{x_3 \to f_C}(x_3) m_{x_5 \to f_C}(x_5),$$

$$m_{f_C \to x_3}(x_3) = \sum_{x_4} \sum_{x_5} f_C(x_3, x_4, x_5) m_{x_4 \to f_C}(x_4) m_{x_5 \to f_C}(x_5),$$

$$m_{x_4 \to f_B}(x_4) = m_{f_C \to x_4}(x_4),$$

$$m_{x_3 \to f_A}(x_3) = m_{f_C \to x_3}(x_3),$$

$$m_{f_B \to x_2}(x_2) = \sum_{x_4} f_B(x_2, x_4) m_{x_4 \to f_B}(x_4),$$

$$m_{f_A \to x_1}(x_1) = \sum_{x_3} f_A(x_1, x_3) m_{x_3 \to f_B}(x_3).$$

The flow of messages is shown in Figure 8.12.

After messages have passed across each edge in each direction we can evaluate the marginal distributions. For example, the marginal distribution $P[x_3]$ is given by:

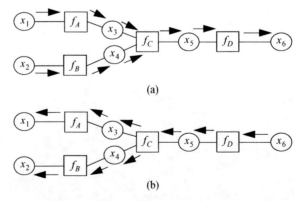

(a)

(b)

Figure 8.12 Message flows. (a) Flow from leaf nodes to root; (b) flow from root to leaf nodes.

$$P[x_3] = \prod_{f \in \aleph(x_3)} m_{f \to x_3}(x_3) = m_{f_A \to x_3}(x_3) m_{f_C \to x_3}(x_3)$$

$$= \left\{ \sum_{x_1} f_A(x_1, x_3) \right\} \left\{ \sum_{x_4} \sum_{x_5} f_C(x_3, x_4, x_5) m_{x_4 \to f_C}(x_4) m_{x_5 \to f_C}(x_5) \right\}$$

$$= \left\{ \sum_{x_1} f_A(x_1, x_3) \right\} \left\{ \sum_{x_4} \sum_{x_5} f_C(x_3, x_4, x_5) \left(\sum_{x_2} f_B(x_2, x_4) \right) \left(\sum_{x_6} f_D(x_5, x_6) \right) \right\}$$

$$= \sum_{x_1} \sum_{x_4} \sum_{x_5} \sum_{x_2} \sum_{x_6} f_A(x_1, x_3) f_C(x_3, x_4, x_5) f_B(x_2, x_4) f_D(x_5, x_6)$$

$$= \sum_{x_1} \sum_{x_4} \sum_{x_5} \sum_{x_2} \sum_{x_6} P[x].$$

8.5.2 The Junction Tree Algorithm

The sum-product algorithm is limited to tree-like graphs. The junction tree algorithm, which is also called the *clique tree algorithm*, is a more general algorithm for arbitrary graphs. A clique tree is a undirected graph whose nodes are cliques. The algorithm transforms the original graph into an undirected singly connected graph, which is the clique tree. The undirected graph is obtained from the multiply connected network by first carrying out moralization of the network and then triangulating the resultant moral graph by finding chordless cycles containing four or more nodes and adding extra edges to eliminate such chordless cycles.

The triangulated graph is then used to construct a new tree-structured undirected graph called a *cluster graph* whose nodes correspond to the *maximal cliques* of the triangulated graph and whose edges connect pairs of cliques that have nodes in common. A maximal clique is a complete subgraph that is not contained in any other complete subgraph. The reason for triangulating the moral graph is to ensure that each node is in a clique with all its parents. From the cluster graph a *cluster tree* is constructed using the *maximum weight spanning tree algorithm*. The weight on each edge is the number of nodes that the two cliques it connects have in common. One method for computing the maximum weight spanning algorithm is the Kruskal's algorithm, which is a modified version of the one discussed in Chapter 7 and works as follows:

1. Sort the edges of the cluster graph in a decreasing order of weight, and let M be the set of edges comprising the maximum weight spanning tree. Set $M = \varnothing$.
2. Add the edge with the largest weight to M, breaking ties arbitrarily.
3. Add the edge with next largest weight to M if and only if it does not form a cycle in M. If there are no remaining edges, exit declare the cluster graph to be disconnected,

4. If M has $n - 1$ edges, where n is the number of vertices in the cluster graph, stop and use M to construct the desired cluster tree; otherwise, go to step 3.

A cluster tree is a junction tree if, for every pair of clusters X and Y, all the nodes on the path from X to Y contain the intersection $X \cap Y$. This condition is called the *junction tree property*. This property ensures that information that is entered at X and is relevant to belief in Y is not forgotten anywhere along the path from X to Y. If $A \in X$ and $A \in Y$ but $A \notin U$, where U is an intermediate cluster, then the information that X has about A cannot propagate to Y. That is, A is forgotten before Y is reached.

In summary, the junction tree algorithm proceeds as follows:

1. Moralize the BN; that is, add undirected edges to *marry* co-parents that are not currently joined. Then drop all directions in the graph to obtain a moral graph.
2. Triangulate the moral graph; that is, add sufficient additional undirected links between nodes such that there are no cycles of length 4 or more distinct nodes.
3. Identify the cliques of the triangulated graph and let them be the nodes of the cluster graph.
4. Use the maximum weight spanning tree algorithm to join the cliques to obtain the cluster tree.
5. Test to see if the junction tree property is obtained; otherwise, construct another cluster tree and repeat the test until a junction tree obtained.

Note that the maximum weight spanning tree is not unique, which is why we test each tree to see if the junction tree property is obtained. A junction tree will be found because of the following theorem:

Theorem 8.1: Every triangulated graph has a junction tree.

As an example of the preceding procedure, we consider the multiply connected BN in Figure 8.13 consisting of nodes A, B, C, D, E, and F.

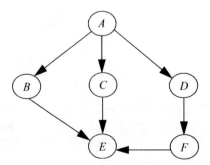

Figure 8.13   Example of a multiply connected BN.

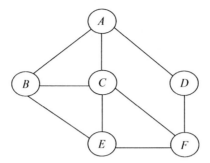

Figure 8.14 Moral graph of Figure 8.13.

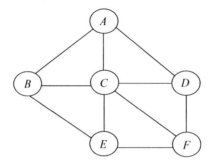

Figure 8.15 Triangulated graph of Figure 8.14.

We proceed as follows to generate the junction tree.

Step 1: Moralizing the Graph

We moralize the graph by observing that nodes B and C are the parents of node E. Similarly, C and F are also the parent of E. Thus, we insert an edge between B and C, insert an edge between C and F, and drop the arrows. The resulting moral graph is shown in Figure 8.14.

Step 2: Triangulating the Graph

We observe that ACFD is a cycle of length 4. Therefore, we use the following elimination algorithm.

1. Begin with a node, say A, and add edges so that all its neighbors are connected to each other
   - If no edges need to be added, then the node cannot be part of a chordless cycle of length greater than 3.
   - As can be seen from Figure 8.15, A belongs to a chordless cycle of length greater than 3. Thus, we add an edge between its neighbors, C and D. Then we remove A from the graph together with all edges connected to it.

2. We continue the process by selecting another node not already eliminated and repeating the process. In this case we choose B and find that it does not belong to a chordless cycle of length greater than 3. Thus, we remove it from the graph together with all edges connected to it.

3. We continue with C and find that it does not belong to a chordless cycle of length greater than 3. Thus, we remove C together with all edges connected to it.

4. We continue with D and find that it does not belong to a chordless cycle of length greater than 3. Thus, we eliminate D together with all edges connected to it.

5. We continue with node E and find that it does not belong to a chordless cycle of length greater than 3. Thus, we remove E from the graph together with all edges connected to it. With this there is no chordless cycle of length greater than 3.

Theorem 8.2: A graph is a triangulated graph if and only if all the nodes can be eliminated without adding any edges.

Thus, according to theorem 8.2, we now have a triangulated graph, which is the moral graph together with the new edges shown in Figure 8.15.

Step 3: Generating the Cluster Graph

We identify the clusters of the triangulated graph as ABC, ACD, BCE, CDF, and CEF. These are the *hypernodes* of the cluster graph with an edge connecting two hypernodes if their intersection is not an empty set. The edge carries a label that is the intersection of the two hypernodes. Thus, if the hypernodes are X and Y, their intersection, which is $X \cap Y$, is called the *separator* of X and Y, and is inserted in a box between hypernodes X and Y. The cluster graph of the triangulated graph of Figure 8.15 is shown in Figure 8.16.

Step 4: Generating the Cluster Tree

The weights of the edges are the cardinalities of the separators. Thus, using the maximum weight spanning tree we generate the cluster tree as follows: We start by selecting separator BC with weight 2 that connects ABC and BCE. Next we choose separator AC with weight 2 that connects ABC and ACD. Next we choose separator CD with weight 2 that connects ACD and CDF. Finally, we choose separator CF with weight 2 that connects CDF and CEF. The cluster tree becomes BCE—ABC—ACD—CDF—CEF as shown in Figure 8.17.

Note that the cluster tree is not unique. We could also obtain the trees CEF—BCE—ABC—ACD—CDF and CDF—CEF—BCE—ABC—ACD. It can be shown that all these cluster trees satisfy the junction tree property. For example, consider the nodes BCE and CDF in Figure 8.17. The intersection {BCE} ∩ {CDF} = C, and C is a member of each node on the path between these two nodes. It can be shown that every pair of nodes in Figure 8.17 satisfies the junction tree property. Thus, Figure 8.17 is a junction tree.

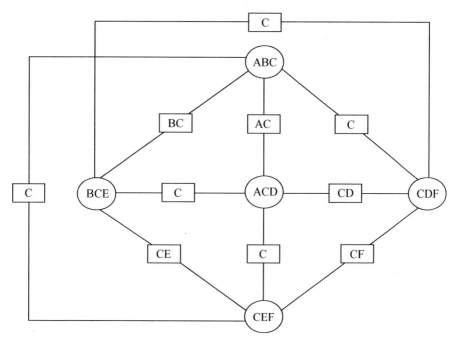

Figure 8.16   Cluster graph of Figure 8.15.

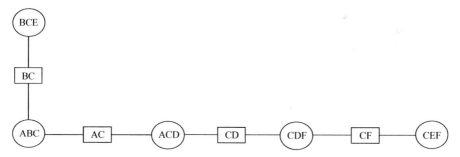

Figure 8.17   Cluster tree of the cluster graph of Figure 8.16.

8.5.2.1 *Belief Propagation in a Junction Tree*   Inference in a directed acyclic graph is based on the conditional independence property of the graph. Since the junction tree is an undirected graph, the concept of *potential function* is used to measure the probability distributions. A potential function is a nonnegative function that maps each instantiation x of a set of random variables X into a nonnegative number called the *potential* of X and denoted by ϕ_X. The number into which ϕ_X maps x is denoted by $\phi_X(x)$.

Potentials do not have a probabilistic interpretation but have the meaning that higher potentials are more probable than lower ones. Let C denote the set of cliques in the graph and let x_c be the set of variables in clique $c \in C$. Let $\phi_c(x_c) \geq 0$ denote the potential function of clique c. Then the joint distribution of the graph is defined by:

$$P[x] = Z^{-1} \prod_{c \in C} \phi_c(x_c),$$

where the parameter Z is a normalizing constant called the *partition function*, which is given by:

$$Z = \sum_x \prod_{c \in C} \phi_c(x_c).$$

The condition $\varphi_c(x_c) \geq 0$ ensures that $P[x] \geq 0$. The potential function for a junction graph satisfies the following conditions:

1. For each clique c and neighboring separator S, the potential of the separator is given by:

$$\phi_S = \sum_{c \setminus S} \phi_c.$$

   When this condition is satisfied for a cluster c and its neighboring separator S, we say that φ_S is *consistent* with φ_c. We say that the graph is *locally consistent* when consistency holds for every cluster-separator pair.

2. For a BN over a set of variables $X = \{X_1, \ldots, X_n\}$, the joint distribution of the belief network is given by:

$$P[X] = \frac{\sum_c \phi_c}{\prod_c \phi_{Sc}},$$

   where Sc is a separator that is adjacent to cluster c.

One important property of junction tree is that for each cluster or separator V, we have that $P[V] = \varphi_V$. Thus, the probability distribution of any variable X_k can be computed using any cluster or separator V that contains X_k as follows:

$$P[X_k] = \sum_{V \setminus \{X_k\}} \phi_V.$$

That is, the summation is over any clique or separator that contains X_k.

Inference in a junction tree deals with how to compute the probability distributions of the variables in the BN. The procedure for doing this is discussed in Huang and Darwiche (1996). Once the junction tree is constructed, the next steps include the following:

- Initialization, during which the junction tree is quantified with belief potentials.
- Global propagation, during which ordered series of messages are passed with the goal of rearranging the potentials so they can become locally consistent.
- Marginalization, during which the probability distributions of the variables of the BN are computed.

8.6   LEARNING BNs

Learning BN deals with how a BN can be learned from available data. Most learning methods combine prior knowledge with data to produce improved knowledge. The process of learning BNs can be classified into two: learning the structure of the network and learning the parameters of the network. Structural learning is the estimation of the topology of the network, and parameter learning is the estimation of the conditional probabilities of the network.

Sometimes the structure of the network is *known* and at other times it is *unknown*. Similarly, the variables can be *observable* (also called *complete data*) or *hidden* (also called *incomplete data*) in all or some of the data points. Thus, there are four classes of learning BNs from data:

- Known structure and observable variables
- Unknown structure and observable variables
- Known structure and hidden variables
- Unknown structure and hidden variables

The techniques used when incomplete data are involved are primarily simulation-based methods. We consider learning BN with complete data.

8.6.1   Parameter Learning

One simple approach to learn the parameters of a BN is to find the parameter set that maximizes the likelihood that the observed data came from the model under study. Two popular methods of doing this are the *maximum likelihood* (MLE) estimation and the *maximum a posteriori* (MAP) estimation. A brief description of these two methods is given.

8.6.1.1   Maximum Likelihood Estimation   Let Θ be a variable that is to be estimated from a set of data X. We know that:

$$P[\Theta|X] = \frac{P[\Theta]P[X|\Theta]}{P[X]} \propto P[\Theta]P[X|\Theta],$$

where the last condition follows from the fact that X is known and, therefore, $P[X]$ is a constant. As we discussed earlier in the chapter, the conditional probability $P[X|\Theta]$ is the *likelihood function*, $P[\Theta|X]$ is the *posterior distribution*, and $P[\Theta]$ is the *prior distribution*. Thus, we have that Posterior Probability $\propto$ Prior Probability $\times$ Likelihood Function.

Consider a family of probability distributions represented by the unknown parameter Θ associated with either a known probability density function (PDF) (for continuous random variables) or a known probability mass function (PMF) (for discrete random variables), denoted by f_Θ. Assume that we draw a sample $\{x_1, x_2, \ldots, x_n\}$ of n values from this distribution, which is an instance of the random sample $\{X_1, X_2, \ldots, X_n\}$, and using f_Θ we compute the multivariate PDF associated with the observed data $f_\Theta(x_1, x_2, \ldots, x_n; \theta)$, where $\theta = \{\theta_1, \theta_2, \ldots, \theta_k\}$ is a sequence of k unknown constant parameters that need to be estimated. If we assume that the data drawn come from a particular distribution that is independent and identically distributed, then we have that:

$$f_\Theta(x_1, x_2, \ldots, x_n; \theta) = f_{\Theta_1}(x_1; \theta) f_{\Theta_2}(x_2; \theta) \ldots f_{\Theta_n}(x_n; \theta) = \prod_{i=1}^{n} f_{\Theta_i}(x_i; \theta).$$

The PDF that specifies the probability of observing data vector $X = \{X_1, X_2, \ldots, X_n\}$ given the parameter θ is $f_{X|\Theta}(x|\theta)$, which is given by:

$$f_{X|\Theta}(x|\theta) = f_{X|\Theta}(x = \{x_1, x_2, \ldots, x_n\}|\theta) = f_{X_1|\Theta}(x_1|\theta) f_{X_2|\Theta}(x_2|\theta) \ldots f_{X_n|\Theta}(x_n|\theta).$$

The problem we are confronted with is the following. Given the observation data and a model of interest, find the PDF (or PMF, in the discrete case) among all PDFs that is most likely to have produced the data. That is, we want to find the most likely $f_{\Theta|X}(\theta|x)$. As we stated earlier, application of Bayes' theorem shows that $f_{\Theta|X}(\theta|x) \propto f_\Theta(\theta)f_{X|\Theta}(x|\theta)$. Thus, to solve this problem, MLE defines the likelihood function $L(\theta|x)$ that reverses the roles of the data vector x and the parameter vector θ; that is,

$$L(\theta|x) = f_{X|\Theta}(x|\theta).$$

Since the likelihood function represents the likelihood of the parameter θ given the observed data, it is a function of θ. Because the data samples are assumed to be independent, we have that:

$$L(\theta|x) = \prod_{i=1}^{n} f_{X_i|\Theta}(x_i|\theta).$$

The maximum likelihood estimate is given by:

$$\hat{\theta} = \arg\max_{\theta} L(\theta|x).$$

The *logarithmic likelihood function* is given by:

$$\Lambda(\theta|x) = \ln\{L(\theta|x)\} = \sum_{i=1}^{n} \ln\{f_{X_i|\Theta}(x_i|\theta)\}.$$

The maximum likelihood estimators of $\theta_1, \theta_2, \ldots, \theta_k$ are obtained by maximizing either $L(\theta|x)$ or $\Lambda(\theta|x)$. Since maximizing $\Lambda(\theta|x)$ is an easier task, the maximum likelihood estimators of $\theta_1, \theta_2, \ldots, \theta_k$ are the simultaneous solutions of the k equations:

$$\frac{\partial \Lambda(\theta|x)}{\partial \theta_i} = 0 \quad i = 1, 2, \ldots, k.$$

As an example, assume that we are given a set of samples $x = \{x_1, x_2, \ldots, x_n\}$ from an exponential distribution. The likelihood function is the joint PDF of the samples at $x_1, x_2, \ldots, x_n$. Let λ be the parameter of the exponential distribution. Then the likelihood function is given by:

$$L(\lambda|x) = \prod_{i=1}^{n} \lambda \exp(-\lambda x_i) = \lambda^n \prod_{i=1}^{n} \exp(-\lambda x_i).$$

The logarithmic likelihood function is given by:

$$\Lambda(\lambda|x) = \ln\{L(\lambda|x)\} = n\ln(\lambda) - \lambda \sum_{i=1}^{n} x_i.$$

Thus, the maximum likelihood estimate is the solution to the equation:

$$\frac{\partial \Lambda(\lambda|x)}{\partial \lambda} = \frac{n}{\lambda} - \sum_{i=1}^{n} x_i = 0.$$

That is,

$$\hat{\lambda} = \frac{n}{\displaystyle\sum_{i=1}^{n} x_i}.$$

As another example, we consider MLE with Gaussian noise. Suppose that $X = \Theta + W$, where Θ and W are independent and $W \sim N(0, \sigma_W^2)$. Then the PDF of W is:

$$f_W(w) = \frac{1}{\sqrt{2\pi\sigma_W^2}} e^{-w^2/2\sigma_W^2}.$$

Given the sample realization x, the likelihood function is given by:

$$f_{X|\Theta}(x|\theta) = f_W(w)\big|_{w=x-\theta} = \frac{1}{\sqrt{2\pi\sigma_W^2}} e^{-(x-\theta)^2/2\sigma_W^2},$$

whose peak value occurs at $\theta = x$. Thus, the maximum likelihood estimate is $\hat{\theta} = x$.

8.6.1.2 Maximum A Posteriori Estimation The MAP estimation is concerned with finding the parameter that maximizes the posterior distribution. That is, given the observation X, find the value of Θ that maximizes $P[\Theta|X]$. Thus, it differs from MLE by the fact that it involves both the likelihood function $P[X|\Theta]$ and the prior distribution $P[\Theta]$ while MLE involves only the likelihood function. Because of this, MAP is sometimes called the *Bayesian estimation*. Thus, assuming that we have a continuous distribution, the MAP estimate is given by:

$$\hat{\theta} = \arg\max_{\theta} f_{\Theta|X}(\theta|x) = \arg\max_{\theta} f_{X|\Theta}(x|\theta) f_{\Theta}(\theta).$$

As an example, we consider the MAP estimation with Gaussian noise. Assume that the observation is $X = \Theta + W$, where $W \sim N(0, \sigma_W^2)$. Because MAP estimation requires knowledge of the prior distribution, we suppose that $\Theta \sim N(\mu_\Theta, \sigma_\Theta^2)$. Thus, we have that:

$$f_{\Theta}(\theta) = \frac{1}{\sqrt{2\pi\sigma_W^2}} e^{-(\theta-\mu_\Theta)^2/2\sigma_\Theta^2},$$

$$f_{X|\Theta}(x|\theta) = f_W(w)\big|_{w=x-\theta} = \frac{1}{\sqrt{2\pi\sigma_W^2}} e^{-(x-\theta)^2/2\sigma_W^2},$$

$$f_{\Theta}(\theta) f_{X|\Theta}(x|\theta) = \frac{1}{2\pi\sigma_\Theta\sigma_W} \exp\left\{ -\frac{(\theta-\mu_\Theta)^2}{2\sigma_\Theta^2} - \frac{(x-\theta)^2}{2\sigma_W^2} \right\}.$$

The MAP estimate maximizes the expression in the last equation, which corresponds to minimizing the term:

$$A = \frac{(\theta-\mu_\Theta)^2}{2\sigma_\Theta^2} + \frac{(x-\theta)^2}{2\sigma_W^2}.$$

That is,

$$\frac{\partial A}{\partial \theta} = \frac{\theta - \mu_\Theta}{\sigma_\Theta^2} - \frac{(x - \theta)}{\sigma_W^2} = 0.$$

Since $\partial^2 A/\partial \theta^2 = \left(1/\sigma_\Theta^2\right) + \left(1/\sigma_W^2\right) > 0$, we solve for θ and obtain the MAP estimate as:

$$\hat{\theta} = \frac{\sigma_W^2 \mu_\Theta}{\sigma_W^2 + \sigma_\Theta^2} + \frac{\sigma_\Theta^2 x}{\sigma_W^2 + \sigma_\Theta^2} = \mu_\Theta + \frac{\sigma_\Theta^2 \{x - \mu_\Theta\}}{\sigma_W^2 + \sigma_\Theta^2}.$$

Note that the solution assumes that the mean μ_Θ and variance σ_Θ^2 are known.

8.6.2 Structure Learning

Since we assume that the variables are observable, the problem in structure learning becomes the following. Given the set of variables in the model, we are to select arcs between them and estimate the parameters. Unfortunately the solution to the problem is usually computationally intractable and sampling methods are often used.

8.7 DYNAMIC BAYESIAN NETWORKS

Dynamic Bayesian networks (DBNs) are an extension of BNs that are designed for dynamic time series data. Thus, the set of random variables $X = \{X_1, \ldots, X_n\}$ that we used in BNs will now be indexed by time as follows: $X^t = \{X_1^t, \ldots, X_n^t\}$, where X_i^t represents the value of the ith random variable at time t. DBNs are time invariant so that the topology is a repeating structure and the CPTs do not change with time.

A DBN consists of a *base network* $X(t)$ that defines the instantaneous state of the system at time t, and a *transition network* R that connects some nodes of $X(t)$ to $X(t + 1)$ for $t = 0, 1, \ldots, T - 1$. Thus, the DBN can be visualized as a BN repeated at each time slice. In networks with first-order Markov property, the parents of a node at time slice t must occur either in time slice t (intraslice arcs) or $t - 1$ (interslice arcs), except for slice $t = 0$ where only intraslice arcs exist. This means that while intraslice arcs can be arbitrary as long as the overall DBN is a DAG, interslice arcs are all from left to right, reflecting the evolution of time. The CPDs within and between time slices are repeated for all $t > 0$. Consequently a DBN can be specified simply by giving two time slices and the arcs between them. The repeating structure and the CPDs are "unrolled" until T. Figure 8.18 shows an example of a DBN unrolled to show T time slices. Note that the transition network is usually assumed to be the same for every pair of neighboring time slices.

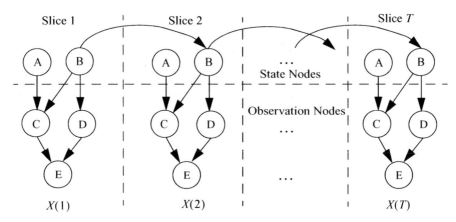

Figure 8.18 Example of a DBN.

The joint probability distribution of the DBN is given by:

$$P[X(t)] = P[X_1(t), X_2(t), \ldots, X_n(t)] = \prod_{t=0}^{T} \prod_{i=1}^{n} P[X_i(t) | Pa(X_i(t))].$$

8.8 PROBLEMS

8.1 Consider the following three expressions for the joint distribution of three random variables X, Y and Z:

   a. $P[X, Y, Z] = P[X|Z]P[Y|Z]P[Z]$

   b. $P[X, Y, Z] = P[X|Z]P[Z|Y]P[Y]$

   c. $P[X, Y, Z] = P[Y|Z]P[Z|X]P[X]$

   Use a DAG to show the graphical representation of each of these expressions.

8.2 Research shows that tuberculosis and lung cancer can each cause dyspnea (or shortness of breath) and a positive chest X-ray. Bronchitis is another cause of dyspnea. A recent visit to a foreign country could increase the probability of tuberculosis. Smoking can cause both cancer and bronchitis. Create a DAG representing the causal relationships among these variables.

8.3 It has been established that smoking causes both bronchitis and lung cancer. Bronchitis and lung cancer both cause fatigue, but only lung cancer can cause a chest X-ray to be positive. If there are no other causal relationships among these variables, create a DAG representing the above causal relationships.

8.4 Consider the DAG shown in Figure 8.19.
 a. Obtain the triangulated graph of the DAG.
 b. Identify the cluster graph of the triangulated graph.

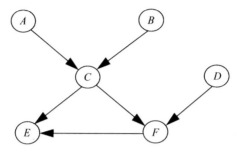

Figure 8.19 Figure for Problem 8.4.

8.5 Consider the function $f(x_1, x_2, x_3) = f_1(x_1, x_2)f_2(x_2, x_3)$.
 a. Show the factor graph for the function.
 b. Compute the messages that are passed.
 c. Compute the marginal distributions $P[x_1]$, $P[x_2]$ and $P[x_3]$.

8.6 Consider the DAG shown in Figure 8.20.
 a. Obtain the triangulated graph of the DAG and identify the maximal cliques.
 b. Obtain the cluster graph of the triangulated graph.
 c. Obtain the junction tree of the cluster graph.

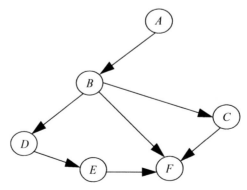

Figure 8.20 Figure for Problem 8.6.

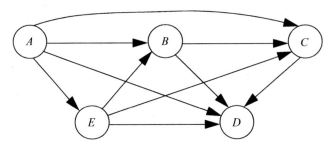

Figure 8.21   Figure for Problem 8.7.

8.7   Consider the DAG shown in Figure 8.21.

    a. Obtain the triangulated graph of the DAG and identify the maximal cliques.

    b. Obtain the cluster graph of the triangulated graph.

    c. Obtain the junction tree of the cluster graph.

9

BOOLEAN NETWORKS

9.1 INTRODUCTION

Genes are either "on" or "off." When a gene is on, it is said to be *expressed*; otherwise, it is not expressed. The state of a gene is called the *expression level* of the gene. It is generally believed that the state of an organism is largely determined by the pattern of expression of its genes. This means that how well a cell performs its function is controlled by the level at which genes are expressed in the cell.

Many biological processes of interest, such as cell differentiation during development, aging, and disease, are controlled by complex interactions over time between hundreds of genes. Also, each gene is involved in multiple functions, which means that understanding the nature of complex biological processes requires determining the spatiotemporal expression patterns of thousands of genes. This also requires understanding the principles that allow biological processes to function in a coherent manner under different environmental conditions.

Boolean networks were introduced as discrete dynamic models originally intended to model genetic regulatory networks (GRNs). They are currently used in modeling other systems including social networks and biological networks because they can monitor the dynamic behavior of large complex systems. They are graphical models whose nodes are Boolean variables that can assume only two values denoted by 0 and 1, which correspond to the logical values "false" and "true," respectively.

Fundamentals of Stochastic Networks, First Edition. Oliver C. Ibe.
© 2011 John Wiley & Sons, Inc. Published 2011 by John Wiley & Sons, Inc.

9.2   INTRODUCTION TO GRNs

Gene expression is the complex process by which cells produce proteins from the instructions encoded into *deoxyribonucleic acid* (DNA). This process includes the synthesis of the corresponding *ribonucleic acid* (RNA), which is called *gene transcription*, and the transformation of the RNA into *messenger ribonucleic acid* (mRNA) that is delivered to places where the mRNA is translated into the corresponding protein chain.

Various activities of cells are controlled by the action of proteins on proteins. After their production, some proteins, called *transcription factors*, then bind back onto the DNA to increase or decrease the expression of other genes. Because proteins are the products of gene expression as well as play a key role in the regulation of gene expression, they significantly contribute to linking genes to each other and forming multiple regulatory circuits in a cell. Thus, a single gene can interact with many other genes in the cell, inhibiting or promoting directly or indirectly, the expression of some of them at the same time.

A gene regulatory network or GRN is a system that deals with the various aspects of the complex interrelationships between genes and their products in a cell. It consists of a set of genes and their mutual regulatory interactions. The interactions arise from the fact that genes code for proteins that may control the expression of other genes, for instance by activating or inhibiting DNA transcription, as discussed earlier.

9.3   BOOLEAN NETWORK BASICS

A Boolean network is a directed graph $G(V, F)$ where $V = \{v_1, v_2, \ldots, v_N\}$ is a set of nodes (genes) such that the nodes $v_i \in V$ are Boolean variables, and $F = \{f_1, f_2, \ldots, f_N\}$ is a set of *Boolean functions*. The nodes usually represent genes and we write $v_i = 1$ to denote that the *i*th node (or gene) is *expressed*, and $v_i = 0$ to denote that it is *not expressed*. Thus, each node follows a switch-like behavior that enables it to move between on and off states with time. The expression level (or expression state) of each gene functionally relates to the expression states of some other genes, using logical rules. Thus, a Boolean function (also called a *predictor*) $f_i(v_{i_1}, v_{i_2}, \ldots, v_{i_k})$ with $k \leq N$ specific input nodes is assigned to node v_i and is used to update its value. The arguments v_{i_j} of f_i are the parents of v_i in G. That is, $v_{i_1}, v_{i_2}, \ldots, v_{i_k}$ are the input nodes to node v_i in $G(V, F)$ and are called the *regulatory set* for v_i. The predictors or Boolean functions are also called *transition rules* because they fix the state transitions for each node in the network.

A predictor is a function of Boolean variables connected by the logic operators such as AND, OR, and NOT. Thus, in a Boolean network, each node is "predicted" by k other nodes by means of a Boolean function, which is the predictor. The predictors are the update rules that are used to define the

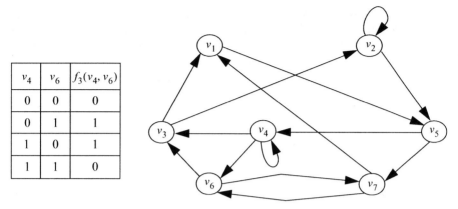

v_4	v_6	$f_3(v_4, v_6)$
0	0	0
0	1	1
1	0	1
1	1	0

Figure 9.1 Example of a Boolean network.

dynamics of each node, and consequently of the whole network. The state of a node is the value of the predictor of the states of its input nodes in the previous time step. Thus, if $v_i(t)$ is the binary value of node v_i at time t, the dynamics of the node is represented by:

$$v_i(t+1) = f_i(v_{i_1}(t), v_{i_2}(t), \ldots, v_{i_k}(t)).$$

We define the vector $v(t) = \{v_1(t), v_2(t), \ldots, v_N(t)\}$ as the *gene activity profile* (GAP) at time t or the global state of the Boolean network at time t. It is assumed that time steps are discrete and in each time step all the nodes are updated at the same time. This is called a *synchronous update*, which is the only type of update that we consider. The predictors together with the regulatory sets determine the *wiring* (i.e., the interconnection of the nodes) of the network.

Figure 9.1 illustrates an example of a Boolean network. There are seven nodes ($N = 7$) and $k = 2$. The predictor for node v_3 is given, which can be seen to be implementing the exclusive OR (XOR) operation. The predictors for the other nodes can be similarly defined.

In the classical Boolean network (or the Kauffman model), the number of inputs per node is fixed to some constant K and the predictors for all nodes in the network are chosen at random from the set of 2^{2^K} possible predictors of K elements, using a specified probability distribution of these functions. Thus, the topology of the network is specified by the number of nodes N in the network and the number of inputs per node K. Note that the number of inputs at a node can include a self-loop. That is, self-loops are permitted and count as inputs to the nodes.

Because of their convenient and easy-to-understand structure, Boolean networks have been used extensively as models for GRNs and other complex networks. Originally introduced by Kauffman (1969), the Boolean network model is particularly applicable to any situation in which the activity of the nodes of the network can be quantized to only two states, ON and OFF, and each node updates its state based on logical relationships with other nodes of the network. Although a Boolean network model may represent a very simplified view of a network, it generally retains meaningful information that can be used to study the dynamics of the network and make inferences regarding the real system it models.

9.4   RANDOM BOOLEAN NETWORKS

In our discussion so far, it has been assumed that the configuration of the network is known. However, if we randomly initialize the network and choose the interconnections of the nodes and the predictors at random, we obtain a random Boolean network (RBN). The states of $G(V, F)$ are the vectors $v(t) = \{v_1(t), v_2(t), \ldots, v_N(t)\}$ that correspond to the expression patterns of V. As discussed earlier, the state vector $v(t)$ is the GAP at time t.

RBNs are obtained by lifting the assumption that each node has the same transition rule and neighborhood. Because of their nonlocal connections and heterogeneous rules, RBNs are better able to model parallel processing systems in biological systems.

The set of all 2^N possible states forms the state space of the network. The transitions of all states together correspond to a state transition of the network from state $S(t)$ to state $S(t + 1)$. A series of state transitions is called a *trajectory*. Because the state space is finite and the trajectory of the network is deterministic, a series of repeating states called an *attractor* will eventually emerge. Thus, all trajectories are periodic, which follows from the fact that as soon as one state is visited a second time, the trajectory will take exactly the same path as the first time. In the language of dynamical systems, an attractor is a set of states to which the system evolves after a long enough time, which is a *fixed point* of the system that captures its long-term behavior.

In RBNs, there are states that lead to an attractor without actually being a part of it. These nonattractor states are called *transient states*, and the set of all transient states is called the *basin of attraction* of the attractor. Thus, a basin of attraction for an attractor consists of the transient states through which the network passes before it enters an attractor. The state space can be divided into a number of periodic attractors and their associated basins of attraction. Because of the deterministic dynamics of the network, none of the basins can overlap. Transient states are visited at most once on any network trajectory.

The number of elements in an attractor is called the *period* of the attractor. An attractor with period 1 is called a *singleton* attractor. Attractors are cyclical and can consist of more than one state. The *cycle length* of an attractor is the number of transitions it takes to return to a particular state on the attractor after leaving the state. We define the *level* of a state as the number of transitions required for the network to transition from the state into an attractor cycle.

9.5 STATE TRANSITION DIAGRAM

One way to analyze Boolean networks is by constructing a truth table that contains the next value of each node. Consider the Boolean network shown in Figure 9.2 and assume that the Boolean functions are defined as follows:

$$v_1(t+1) = f_1(v_2) = v_2(t),$$
$$v_2(t+1) = f_2(v_1, v_3) = v_1(t) \cap \overline{v_3(t)},$$
$$v_3(t+1) = f_3(v_1, v_2) = v_1(t) \cap \overline{v_2(t)}.$$

The truth table of the network is shown in Figure 9.3. Associated with the truth table is the state transition diagram. The truth table lists all possible combinations of inputs and the corresponding output of each Boolean function, and the state transition diagram represents transitions between the states defined by the outputs of the truth table.

From Figure 9.3b we find that there are three attractors that depend on the initial state of the network. For example, if the initial state is 110, the network never leaves that state. Similarly, if the initial state is 000, the network does not leave that state.

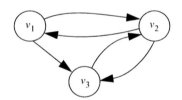

Figure 9.2 Boolean network example.

Current State			Next State		
$v_1(t)$	$v_2(t)$	$v_3(t)$	$v_1(t+1)$	$v_2(t+1)$	$v_3(t+1)$
0	0	0	0	0	0
0	0	1	0	0	0
0	1	0	1	0	0
0	1	1	1	0	0
1	0	0	0	1	1
1	0	1	0	0	1
1	1	0	1	1	0
1	1	1	1	0	0

(a)

(b)

Figure 9.3 (a) Truth table and (b) state transition diagram of Figure 9.2.

9.6 BEHAVIOR OF BOOLEAN NETWORKS

Two of the factors that have an impact on the mode of operation of Boolean networks are the connectivity of the network and the function bias. Let k_i denote the number of inputs to node i; alternatively, k_i is the indegree of node i. Then the connectivity of the network, K, is defined by:

$$K = \frac{1}{N} \sum_{i=1}^{N} k_i.$$

That is, K is the mean number of input variables used by a predictor in the network. An $N\text{-}K$ Boolean network is a network with N nodes such that the indegree of each node is K. Here $k_i = K$ for all i.

The bias p of a predictor is the probability that the function takes the value 1. If $p = 0.5$, the function is said to be unbiased. RBNs exhibit three distinct phases of dynamics, depending on the parameters K and p. These phases, which are identified by the ways in which perturbations (such as changing the state of a node) spread through the network, are:

- Ordered or frozen
- Chaotic
- Critical

At relatively low connectivity (i.e., low indegree, usually $K < 2$), the network is in an ordered state, which is characterized by high stability of states; perturbations die out with short transients and different states tend to converge. Similarly, at relatively high connectivity, usually $K \geq 3$, the network is in a chaotic state, which is characterized by a low stability of the states where perturbations grow exponentially, causing extensive changes to the network state, and different states tend to diverge. The critical phase is the "edge of chaos" between the preceding two regimes. Here $K = 2$, and perturbations continue to affect a limited number of nodes causing the system to be marginally stable. The length of a period is proportional to $\sqrt{N}$, where N is the number of nodes in the network.

Phase transition can also be observed by biasing the predictors f_i so that the output variables switch more or less frequently if the input variables are changed. The "network bias" p represents the fraction of 1's or 0's in the output, whichever is the majority for that function. In general, on increasing p at fixed K, a phase transition is observed from chaotic to frozen behavior. For $K < 2$, the unbiased, random value happens to fall above the transition in the frozen phase, while for $K \geq 3$ the opposite occurs.

Phase transition is typically quantified by using a measure of sensitivity to initial conditions of damage spreading. Specifically, an initial state A of the network is defined. Then the state of a single node is perturbed by changing a 0 to 1 or vice versa to produce the network B. Both networks are then run for many time steps and monitored. If the damage has spread so that the network state now differs from the original activity pattern in an increasing number of places, the network is called *chaotic*, which is characterized by sensitive dependence on initial conditions. If the perturbation has disappeared so that both networks display the same pattern, the network is called *frozen* (or *ordered*). If the states of the two networks still differ in a limited number of places so that the damage may not have spread or disappeared, the network is said to display *critical* behavior.

Quantitatively we define the Hamming distance between A and B by:

$$D(A, B) = \frac{1}{N} \sum_{i=1}^{N} |a_i - b_i|.$$

Let the convergence/divergence parameter δ be defined as follows:

$$\delta = D(A, B)_{t \to \infty} - D(A, B)_{t=0}.$$

Thus, δ is the difference between the normalized distance between the final states and the normalized distance between the initial states. Observe that $D(A, B)_{t=0} = 1/N$ since we started with B differing from A in only one bit. If $\delta < 0$, there is convergence as both initial states tend to the same attractor and the network is stable. Similarly, if $\delta > 0$, then the dynamics of similar initial states diverge, which is a characteristic of the chaotic phase. As p is changed the level of connectivity corresponding to the critical phase, K_c, is given by:

$$K_c = \frac{1}{2p(1-p)}.$$

This is shown in the graph in Figure 9.4.

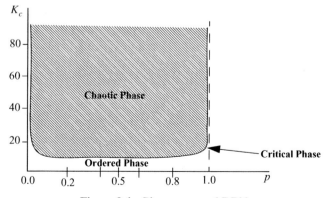

Figure 9.4 Phase space of RBN.

9.7   PETRI NET ANALYSIS OF BOOLEAN NETWORKS

Petri nets (PNs) can be used to analyze the dynamics of a Boolean network. Specifically, they can be used to model the interactions that take place at each node. That is, they can be used to implement the different Boolean functions associated with the nodes in a Boolean network. In this section we give a brief introduction to PNs, including extensions to the original or basic PN. We then discuss how PNs can be used to model Boolean networks.

9.7.1   Introduction to PNs

PNs were introduced by Carl Petri in the early 1960s as a mathematical tool for modeling distributed systems and, in particular, notions of concurrency, nondeterminism, communication, and synchronization. They have been successfully used for concurrent and parallel systems modeling and analysis, communication protocols, performance evaluation, and fault-tolerant systems.

The PN provides a means of representing the structure of the system by modeling system states. The system is represented by a finite set of *places* $P = \{P_1, P_2, \ldots, P_n\}$; a finite set of *transitions* $T = \{t_1, t_2, \ldots, t_m\}$; an input function $I: T \rightarrow P$ that maps a transition to a collection of input places; and an output function $O: T \rightarrow P$ that maps a transition to a collection of output places. A PN can be considered a bipartite graph with places, which are represented by circles, connected by arcs to transitions that are represented by bars, and transitions connected by arcs to places. The set of arcs is given by $A \subseteq \{P \times T\} \cup \{T \times P\}$. Thus, we define two types of arcs:

- An input arc that connects a place to a transition
- An output arc that connects a transition to a place

The state of a PN is represented by the distribution of *tokens* in the places. This distribution of tokens is called a *marking*. The state space of a PN is therefore the set of all possible markings. The initial distribution of tokens is called the *initial marking* of the PN and is defined by $M_0 = \{m_1^0, m_2^0, \ldots, m_n^0\}$, where $m_j^0 = 0, 1, 2, \ldots; j = 1, 2, \ldots, n$. The marking can be regarded as a mapping from the set of places to the natural numbers; that is, $M: P \rightarrow \aleph$, where $M(P_i) = m_i, i = 1, 2, \ldots, n$. Thus, we may define a PN as a four-tuple:

$$PN = (P, T, A, M_0).$$

A place represents a condition that must be met before an event can take place. The placement of a token in a place means that the condition associated with the place has been met. Similarly, a transition represents an event. The dynamic nature of the PN is derived from its execution (or the movement of tokens within the PN). Before an execution can take place, a transition must be *enabled*, which means that each input place of the transition must have at

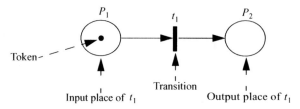

Figure 9.5   Components of a PN.

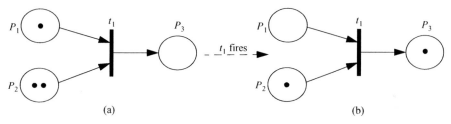

Figure 9.6   Example of PN markings. (a) $M_0 = (1, 2, 0)$; (b) $M_1 = (0, 1, 1)$.

least one token. An enabled transition may *fire*, which indicates that the event may occur. When this happens, a token is taken from each input place of the transition and a token is deposited in each output place of the transition, resulting in a new system state (or a new marking). The new state may enable one or more transitions, and the transitions may continue to fire as long as at least one of them is enabled. Figure 9.5 shows the components of a PN.

As an example of a transition firing, consider the PN shown in Figure 9.6a that consists of one transition and three places. Places P_1 and P_2 are input places to transition t_1 and place P_3 is the output place of t_1. The initial marking is $M_0 = (1, 2, 0)$. Currently t_1 is enabled, and when it fires we have the marking $M_1 = (0, 1, 1)$ as shown in Figure 9.6b.

Sometimes a transition needs more than one token from a given input place to be enabled. In this case we say that the *multiplicity* of that place is the number of required tokens. When the transition fires, it takes the number of tokens equal to the multiplicity of the place from that input place. It is also possible for an output place to have a multiplicity of more than one, which means that more than one token will be deposited in the place when the transition fires. An example is shown in Figure 9.7 where input place P_2 has a multiplicity of 2.

In some applications a transition may be inhibited from firing under certain conditions. This feature is captured by defining an input *inhibitor arc* that has a rounded rather than an arrow-pointed head, as shown in Figure 9.8. Here, transition t_1 is not enabled because place P_1 has a token in it; if it did not, t_1 would be enabled.

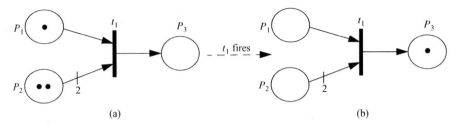

Figure 9.7 PN with multiplicity of $P_2 = 2$. (a) $M_0 = (1, 2, 0)$; (b) $M_1 = (0, 0, 1)$.

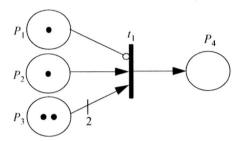

Figure 9.8 PN with inhibitor arc.

The original PN did not support time function. But application of PNs to a real system requires some time scheme to model delay characteristics in the system behavior. Thus, extensions to the original PN have been proposed to permit more flexibility in PN modeling. These extensions include the following:

- *Timed PNs*, which are PNs to which a time delay is assigned to the interval between the instant a transition is enabled and when the transition fires. The introduction of time delays is particularly useful for performance evaluation and scheduling problems of dynamic systems.
- *Stochastic PNs*, which are timed PNs in which the time delays are probabilistically specified. In particular, the firing times are usually assumed to be exponentially distributed.
- *Generalized stochastic PNs*, which are stochastic PNs that have two types of transitions: *immediate transitions* that fire as soon as they are enabled, and *timed transitions* that fire after an exponentially distributed time when they are enabled.
- *Continuous PNs*: The PNs described earlier may be referred to as *discrete* PNs. A continuous PN uses places and transitions in a manner that is different from the way a discrete PN uses them. In a continuous PN, a place contains a real value while in a discrete PN a place contains an integer value, which is the number of tokens. Similarly, in a continuous PN, a transition has a *firing quantity*, which is a real number that repre-

sents the amount of resource that passes through the transition simultaneously when it fires.

- *Hybrid PNs*, which are PNs that are combinations of the discrete PN and the continuous PN.

9.7.2 Behavioral Properties of PNs

As stated earlier, PNs were used for modeling the notions of concurrency, nondeterminism, communication, and synchronization in distributed systems. Consider the PN shown in Figure 9.9 where transitions t_1 and t_2 are enabled and the other transitions are not enabled. Since they do not share any resource (i.e., they do not have a common input place), they can fire at the same time. Thus, we have a parallel or concurrent operation.

When transition t_1 fires, it deposits a token in place P_3. Similarly, when transition t_2 fires, it deposits a token in place P_4. When places P_3 and P_4 contain a token each, transition t_3 is enabled since both places are the input places of the transition. If only one of them contains a token, the transition is not enabled. Thus, permitting more than one place to be the input place of a transition enables us to model synchronization whereby a transition cannot fire until each input place has at least one token. This is illustrated in Figure 9.10 where t_3 is enabled.

When transition t_3 fires, it deposits a token in place P_5, and this causes transitions t_4 and t_5 to be simultaneously enabled, as shown in Figure 9.11. However, only one of them can fire because if t_4 fires, then t_5 is disabled; similarly, if t_5

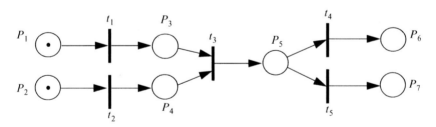

Figure 9.9 Modeling concurrency/parallelism.

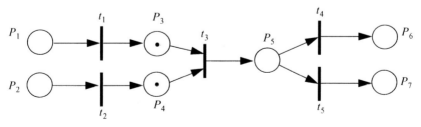

Figure 9.10 Modeling synchronization.

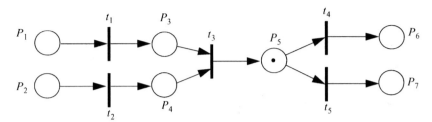

Figure 9.11 Modeling mutual exclusion.

fires, then t_4 is disabled. In this way we can model mutual exclusion or conflict.

Recall that a marking of a PN defines the distribution of tokens in the PN. A marking M_j is said to be *reachable* from a marking M_i if, starting from M_i there is a sequence of transitions whose firings generate M_j. This reachability feature is used to find erroneous states. A PN can be analyzed by constructing its *reachability graph*, which captures the possible firing sequences that can result from a given initial marking.

Another property of a PN is *liveness*, which is the property that the PN never produces a marking in which no transition is enabled. It is used to check for the potential for *deadlock* in a system. Deadlock is the condition in which one component of a system is waiting for another component of the system to release the resources it needs to complete some transaction, and the other component is also waiting for it to release the resources it also needs to complete its own transaction.

Boundedness is a property that ensures that the number of tokens in any place cannot exceed some predefined value. A k-bounded PN has the property that the maximum number of tokens in any place is k. It is used in some systems to estimate the capacity of the components of the system. A 1-bounded PN is referred to as a *safe* PN. Safeness is an important property because it means that the PN has a restricted state space that is amenable to automatic analysis.

Finally, *conservation* is a property that ensures that the number of tokens in every marking is constant; that is, tokens are neither created nor destroyed. The property is used in many resource allocation applications.

Stochastic PNs have been used to model polling systems in Ibe and Trivedi (1990), vacation queueing systems in Ibe and Trivedi (1991), and client–server systems in Ibe et al. (1993).

9.7.3 PN Model of Boolean Networks

Consider the Boolean network in Figure 9.1 where the predictor for node v_3 is given. The PN for this predictor is given in Figure 9.12. Places $P_1, P_2, \ldots, P_6$ represent nodes $v_1, v_2, \ldots, v_5$ and v_6, respectively. A token arriving at place P_k

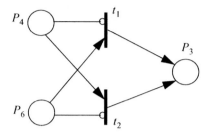

Figure 9.12   PN for the predictor of node v_3 of Figure 9.1.

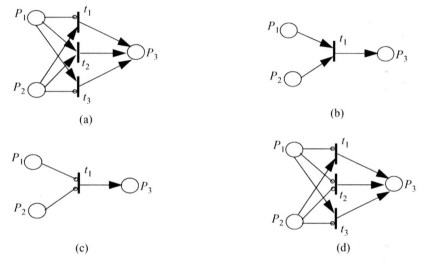

Figure 9.13   PNs for different Boolean functions. (a) OR; (b) AND; (c) NOR; (d) NAND.

indicates that cell k is expressed. Transitions t_1 and t_2 are used to implement the different predictors. Specifically, t_1 may fire when there is no token in P_4 and there is a token in P_6, which implements the condition $v_4 = 0$ and $v_6 = 1$ corresponding to the second row of the table in Figure 9.1. Similarly, t_2 may fire when there is a token in P_4 and there is no token in P_6, which implements the third row. Observe that cell 3 is not expressed when cells 4 and 6 are expressed, which corresponds to a token in P_4 and a token in P_6; thus, we enforce the last row of the table in Figure 9.1. Observe also that the first row is enforced since there can be no token in P_3 when there is no token in P_4 and no token in P_6; in other words, cell 3 cannot be expressed when both cells 4 and 6 are not expressed.

Figure 9.13 shows the PNs for the OR, AND, NOR, and NAND logic functions when the input consists of two Boolean variables whose states are represented by P_1 and P_2, respectively.

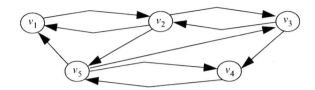

Figure 9.14   Simple Boolean network.

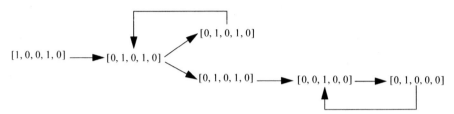

Figure 9.15   State transition diagram of the simple Boolean network.

To put everything in perspective, consider the simple Booloean network shown in Figure 9.14 that consists of five nodes labeled $v_1, v_2, \ldots, v_5$. Assume that the predictors are as follows:

$$f_1(v_2, v_5) = \text{AND}(v_2, v_5),$$
$$f_2(v_1, v_3) = \text{OR}(v_1, v_3),$$
$$f_3(v_2, v_5) = \text{XOR}(v_2, v_5),$$
$$f_4(v_3, v_5) = \text{AND}(v_3, v_5),$$
$$f_5(v_2, v_4) = \text{AND}(v_2, v_4),$$

where we have used the notation $\text{AND}(v_2, v_5) = v_2 \cap v_5$, $\text{OR}(v_2, v_5) = v_2 \cup v_5$, and so on.

Let place P_k represent the state of node k, $k = 1, \ldots, 5$. Assume that cells 1 and 4, which are represented by node 1 and node 4, respectively, are initially expressed. That is, $v(0) = [1, 0, 0, 1, 0]$, and the GAP changes according to the state transition diagram shown in Figure 9.15.

An alternative way to analyze the dynamics of the Boolean network is to use the PN. The fact that cells 1 and 4 are initially expressed means that P_1 and P_4 initially contain a token each. Figure 9.16 is the PN for the Boolean network of Figure 9.14.

Figure 9.17 shows the reachability diagram of the PN, which is obtained as follows. The initial marking is $M_0 = 10010$. Under this marking, only one transition, t_4, is enabled. When it fires, it generates the marking $M_1 = 01010$. Under this marking, two transitions, t_6 and t_8, are enabled. If t_6 fires, the marking $M_2 = 00110$ is generated, which causes t_2 to be enabled and its firing generates $M_1 = 01010$. Similarly, if t_8 fires, the marking $M_3 = 00001$ is generated, and this causes transition t_5 to be enabled. When transition t_5 fires, the marking

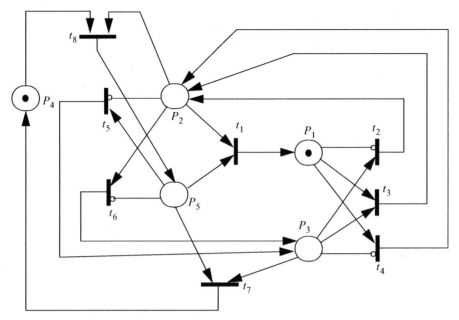

Figure 9.16 PN model of the simple Boolean network.

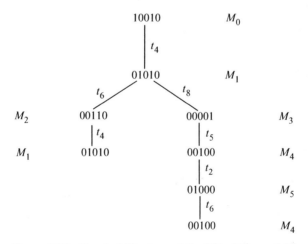

Figure 9.17 Reachability tree of the PN of Figure 9.16.

$M_4 = 00100$ is obtained, which causes transition t_2 to be enabled. When t_2 fires, the marking $M_5 = 01000$ is obtained, which causes transition t_6 to be enabled and its firing produces the marking $M_4 = 00100$. The tree is pruned when a previously obtained marking is obtained again.

The final aspect of the model is the reachability graph, which is a directed graph that is obtained from the reachability tree as follows. The nodes of the

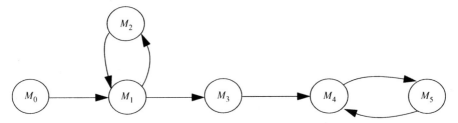

Figure 9.18   Reachability graph of the reachability tree of Figure 8.17.

graph are the markings, and a link exists between node M_i and node M_k if
there is an enabled transition in M_i whose firing generates M_k, which means
that M_k is reachable from M_i in a single transition. The reachability graph of
the reachability tree in Figure 9.17 is shown in Figure 9.18, which is similar to
the state transition diagram shown in Figure 9.15.

From the reachability graph we can see that the initial marking is never
visited again. Other observations include the fact that on the long run the
system oscillates between M_4 and M_5. Thus, all the other markings are transient
states. This means that M_4 and M_5 are the attractor of the network while M_0,
M_1, M_2 and M_3 are transient states that constitute the basin of attraction of
the attractor. Thus, one of the strengths of the PN analysis is that it enables us
to identify the attractor and the basin of attraction. Also, if we assume that
the firing times are governed by some probabilistic law, the reachability graph
can enable us to compute some performance parameters, such as the mean
time to enter an attractor state. For example, if we assume that the firing times
are exponentially distributed, then the reachability graph is a continuous-time
Markov chain, and the mean time to enter an attractor state is the mean time
to absorption of the Markov chain.

9.8   PROBABILISTIC BOOLEAN NETWORKS

One of the drawbacks of Boolean networks in modeling GRNs is their inher-
ent deterministic nature, which is linked to the Boolean logic. As stated earlier,
in a Boolean network, each target gene is "predicted" by several other genes
by means of a Boolean function (or a predictor). Thus, the dynamics of a
Boolean network are deterministic because the system always follows a static
transition mechanism regulated by the Boolean functions and ultimately
settles in the attractor state from where it cannot move.

Probabilistic Boolean networks (PBNs) are designed to overcome the
deterministic rigidity of Boolean networks. They share the appealing properties of Boolean networks, but are able to cope with uncertainty both in the
data and in the model selection. The basic idea of a PBN is to define more
than one Boolean function for each node, thereby permitting the model to
have more flexibility. Thus, in a PBN a number of Boolean functions $f_k^{(i)}, k = 1,$

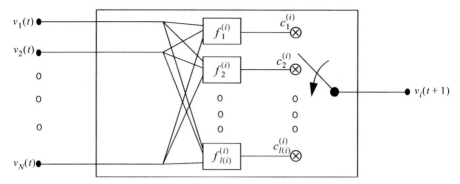

Figure 9.19 Basic building block of a PBN.

$2, \ldots, l(i)$, is associated with each vertex v_i, instead of only one Boolean function as in the RBN, where $l(i) \leq 2^{2^N}$ is the total number of possible Boolean functions of gene i available. This means that there are L different possible realizations of a PBN, and each realization represents a Boolean network, where L is defined by:

$$L = \prod_{i=1}^{N} l(i).$$

The probability of choosing the Boolean function $f_k^{(i)}$ is $c_k^{(i)}$, where $0 \leq c_k^{(i)} \leq 1$ and:

$$\sum_{k=1}^{l(i)} c_k^{(i)} = 1 \quad i = 1, 2, \ldots, N.$$

The probability $c_k^{(i)}$ represents the confidence in using the Boolean function $f_k^{(i)}$ to explain the dynamics of v_i and can be estimated by using a statistical method. Figure 9.19 illustrates the basic building block of a PBN.

9.9   DYNAMICS OF A PBN

The dynamics of the PBN are essentially the same as for Boolean networks, but at any given point in time, the value of each node is determined by one of the possible predictors, chosen according to its corresponding probability. This can be interpreted by saying that at any point in time, we have one out of L possible networks or realizations. Also, since a PBN is a finite collection of Boolean networks, its attractor cycles are the attractor cycles of its constituent Boolean networks.

The dynamic behavior of a PBN is represented by a Markov chain whose transition matrix is determined by the Boolean functions and their probabilities of occurrence. As discussed earlier,

$$P\left[v_i(t+1) = f_k^{(i)}(t)\right] = c_k^{(i)}$$

is the probability that the predictor $f_k^{(i)}$ is chosen to update node v_i. The jth possible realization of the network, f_j, is defined by:

$$f_j = \left\{f_{j_1}^{(1)}, f_{j_2}^{(2)}, \ldots, f_{j_N}^{(N)}\right\} \quad 1 \le j_i \le l(i), \quad i = 1, 2, \ldots, N.$$

The probability of choosing a Boolean function independently for each node (i.e., choosing an independent PBN) is given by:

$$p_j = \prod_{i=1}^{N} c_{j_i}^{(i)}, \quad j = 1, 2, \ldots, L,$$

where L is the maximum possible number of different realizations of Boolean networks, as defined earlier. We can describe the dynamics of a PBN by the state transition probabilities. Let $v(t) = m$ be the GAP of a PBN at time t and let $v(t + 1) = n$ be the GAP at time $t + 1$. The state transition probability is given by:

$$p_{mn} = P\left[v(t+1) = n \mid v(t) = m\right].$$

Because there are 2^N possible GAPs, we can represent the state transition probabilities by a $2^N \times 2^N$ transition probability matrix. From this matrix we can determine the steady-state behavior of the PBN.

9.10 ADVANTAGES AND DISADVANTAGES OF BOOLEAN NETWORKS

The estimation of gene regulatory networks using the Boolean network offers several advantages. These include the fact that the Boolean network model effectively explains the dynamic behavior of living systems. A simplistic Boolean formalism can represent realistic complex biological phenomena such as cellular state dynamics that exhibit switch-like behavior, stability, and hysteresis. It also enables the modeling of nonlinear relations in complex living systems. Also, Boolean algebra is an established science that provides a large set of algorithms that are already available for supervised learning in the binary domain, such as the logical analysis of data, and Boolean-based classification algorithms. Finally, the fact that the variables are Boolean and thus have binary values improves the accuracy of classification and simplifies the obtained models by reducing the noise level in experimental data.

One of the major drawbacks of Boolean networks is that they require extremely high computing times to construct reliable network structures. Consequently, most Boolean network algorithms can be used only with a small number of genes and a low indegree value. The associated high computing times are a major problem in the study of large-scale gene regulatory and gene interaction systems using Boolean networks.

9.11 PROBLEMS

9.1 Consider a Boolean network with three nodes v_1, v_2, and v_3 shown in Figure 9.20. Construct the truth table of the network and draw the state transition diagram.

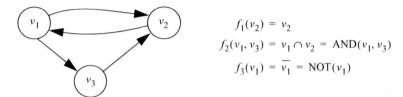

$$f_1(v_2) = v_2$$
$$f_2(v_1, v_3) = v_1 \cap v_2 = \text{AND}(v_1, v_3)$$
$$f_3(v_1) = \overline{v_1} = \text{NOT}(v_1)$$

Figure 9.20   Figure for Problem 9.1.

9.2 Consider the Boolean network shown in Figure 9.21.
Obtain the following:
a. The PN model of the network
b. The reachability tree of the PN
c. The reachability graph of the reachability tree
d. The attractor A_1 of the network

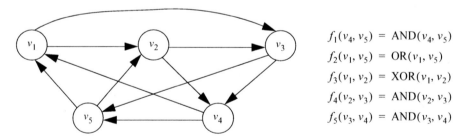

$$f_1(v_4, v_5) = \text{AND}(v_4, v_5)$$
$$f_2(v_1, v_5) = \text{OR}(v_1, v_5)$$
$$f_3(v_1, v_2) = \text{XOR}(v_1, v_2)$$
$$f_4(v_2, v_3) = \text{AND}(v_2, v_3)$$
$$f_5(v_3, v_4) = \text{AND}(v_3, v_4)$$

Figure 9.21   Figure for Problem 9.2.

9.3 Consider the Boolean network shown in Figure 9.22.
Obtain the following:
a. The PN model of the network
b. The reachability tree of the PN
c. The reachability graph of the reachability tree
d. The attractor A_2 of the network

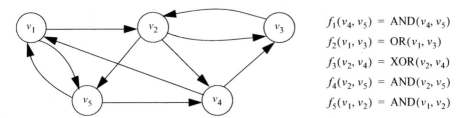

$$f_1(v_4, v_5) = \text{AND}(v_4, v_5)$$
$$f_2(v_1, v_3) = \text{OR}(v_1, v_3)$$
$$f_3(v_2, v_4) = \text{XOR}(v_2, v_4)$$
$$f_4(v_2, v_5) = \text{AND}(v_2, v_5)$$
$$f_5(v_1, v_2) = \text{AND}(v_1, v_2)$$

Figure 9.22   Figure for Problem 9.3.

9.4   Consider the Boolean network shown in Figure 9.23.
Obtain the following:
a. The PN model of the network
b. The reachability tree of the PN
c. The reachability graph of the reachability tree
d. The attractor A_3 of the network

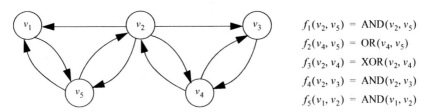

$$f_1(v_2, v_5) = \text{AND}(v_2, v_5)$$
$$f_2(v_4, v_5) = \text{OR}(v_4, v_5)$$
$$f_3(v_2, v_4) = \text{XOR}(v_2, v_4)$$
$$f_4(v_2, v_3) = \text{AND}(v_2, v_3)$$
$$f_5(v_1, v_2) = \text{AND}(v_1, v_2)$$

Figure 9.23   Figure for Problem 9.4.

9.5   Consider the Boolean networks in Figures 9.21–9.23, which we refer to by their attractors as A_1, A_2, and A_3, respectively. Since they are all derived from the same set of nodes, assume that they are selected in a probabilistic manner according to the following transition probabilities: Given that the network is currently in A_1, it will enter A_1 again with probability 0.4, it will enter A_2 next with probability 0.5, and it will enter A_3 next with probability 0.1. Similarly, given that it is currently in A_2, it will enter A_1 with probability 0.3, it will enter A_2 again with probability 0.3, and it will enter A_3 next with probability 0.4. Finally, given that it is currently in A_3, it will enter A_1 with probability 0.3, it will enter A_2 next with probability 0.2, and it will enter A_3 again with probability 0.5. Determine the limiting state probabilities of the attractors of the network.

10

RANDOM NETWORKS

10.1 INTRODUCTION

Many natural and social systems are usually classified as complex due to the interwoven web through which their constituents interact with each other. Such systems can be modeled as networks whose vertices denote the basic constituents of the system and the edges describe the relationships among these constituents. A few examples of these *complex networks* are given in Newman (2003) and include the following:

a. *Social Networks*: In a social network, the vertices are individuals or groups of people and the edges represent the pattern of contacts or interactions between them. These interactions can be friendships between individuals, business relationships between companies, or intermarriages between families.

b. *Citation Networks*: A citation network is a network of citations between scientific papers. Here, the vertices represent scientific papers and a directed edge from vertex v to vertex u indicates that v cites u.

c. *Communication Networks*: An example of a communication network is the World Wide Web (WWW) where the vertices represent home pages and directed edges represent hyperlinks from one page to another.

Fundamentals of Stochastic Networks, First Edition. Oliver C. Ibe.
© 2011 John Wiley & Sons, Inc. Published 2011 by John Wiley & Sons, Inc.

d. *Biological Networks*: A number of biological systems can be represented as networks. Examples of biological networks that have been extensively studied are neural networks, metabolic reaction networks, protein interaction networks, genetic regulatory networks, and food networks.

Complex networks are an emerging branch of random graph theory and have attracted the attention of physical, social, and biological scientists in recent years. Thus, one may define complex networks as networks that have more complex architectures than random graphs. As mentioned earlier, these networks have attracted much attention recently, and many research papers and texts have been published on the subject. Extensive discussion on complex networks can be found in the following texts: Durrett (2007), Vega-Redondo (2007), Barrat et al. (2008), Jackson (2008), Lewis (2009), Easley and Kleinberg (2010), Newman (2010), and van Steen (2010). Also, the following review papers are highly recommended: Albert and Barabasi (2002), Dorogovtsev and Mendes (2002), Newman (2002), Boccaletti et al. (2006), Costa et al. (2007), and Dorogovtsev et al. (2008).

10.2   CHARACTERIZATION OF COMPLEX NETWORKS

Some key measures and characteristics of complex networks are discussed in this section.

These include degree distribution, geodesic distances, centrality measures, clustering, network entropy, and emergence of giant component.

10.2.1   Degree Distribution

As we discussed in Chapter 7, the degree $d(v_i)$ of a vertex v_i is the number of edges attached to it. Let K denote the degree of a vertex and let $p_K(k) = P[K = k] \equiv p_k$ be the probability mass function (PMF) of K. Then p_k specifies, for each k, the fraction of vertices in a network with degree k, which is also the probability that a randomly selected vertex has a degree k, and defines the degree distribution of the network. The most commonly used degree distributions for a network with N vertices include the following:

a. Binomial distribution, whose PMF is given by:

$$p_K(k) = \binom{N-1}{k} p^k (1-p)^{n-k-1} \quad k = 0, 1, \ldots, N-1,$$

where p is the probability that a given vertex has an edge to another given vertex and there are $N-1$ possible edges that could be formed. Each edge is formed independently.

b. Poisson distribution, whose PMF is given by:

$$p_K(k) = \frac{\beta^k}{k!} e^{-\beta} \quad k = 0, 1, \ldots,$$

where β is the average degree. The Poisson distribution is an approximation of the binomial distribution when $N \to \infty$ and $\beta = (N-1)p$ is constant.

c. Power-law distribution, whose PMF is given by:

$$p_K(k) = Ak^{-\gamma} \quad k = 1, 2, \ldots,$$

where $\gamma > 1$ is a parameter that governs the rate at which the probability decays with connectivity; and A is a normalizing constant, which is given by:

$$1/A = \sum_{k=1}^{\infty} k^{-\gamma}.$$

10.2.2 Geodesic Distances

Given two vertices v_i and v_j in a network, we may want to know the geodesic distance $d(v_i, v_j)$ between them; that is, the shortest path between them or the minimum number of edges that can be traversed along some path in the network to connect v_i and v_j. As discussed earlier, the *diameter d_G* of a network is the longest geodesic distance in the network. That is,

$$d_G = \max_{i,j} d(v_i, v_j).$$

The diameter characterizes the ability of two nodes to communicate with each other; the smaller d_G is, the shorter is the expected path between them. In many networks, the diameter and the average distance are on the order of $\log(N)$, which means that as the number of vertices increases, the diameter of the network increases somewhat slowly. Alternatively, the diameter and average distance can be small even when the number N of nodes is large. This phenomenon is often called the *small-world phenomenon*. Small-world networks are discussed later in this chapter. Note that while the geodesic distance between two vertices is unique, the *geodesic path* that is the shortest path between them need not be unique because two or more paths can tie for the shortest path.

10.2.3 Centrality Measures

Several measures have been defined to capture the notion of a node's importance relative to other nodes in a network. These measures are called *centrality measures* and are usually based on network paths. We define two such measures, which are the *closeness centrality* and the *betweenness central-ity*. The closeness centrality of a node v expresses the average geodesic dis-tance between node v and other nodes in the network. Similarly, the betweenness centrality of a node v is the fraction of geodesic paths between other nodes that node v lies on.

Betweenness is a crude measure of the control that node v exerts over the flow of traffic between other nodes. Consider a network with node set V. Let $\sigma_{uv} = \sigma_{vu}$ denote the number of shortest paths from $u \in V$ to $v \in V$. Let $\sigma_{uw}(v)$ denote the number of shortest paths from node u to node w that pass through node v. Then the closeness centrality $c_C(v)$ and the betweenness centrality $c_B(v)$ of node v are defined, respectively, as follows:

$$c_C(v) = \frac{1}{\sum_{w \in V, w \neq v} d(v, w)},$$

$$c_B(v) = \sum_{u \neq v \neq w \in V} \frac{\sigma_{uw}(v)}{\sigma_{uw}}.$$

High closeness centrality scores indicate that a node can reach other nodes on relatively short paths, and high betweenness centrality scores indicate that a node lies on a considerable fraction of the shortest paths connecting other nodes.

10.2.4   Clustering

One feature of real networks is their "cliquishness" or "transitivity," which is the tendency for two neighbors of a vertex to be connected by an edge. Clustering is more significant in social networks than in other networks. In a social network, it is the tendency for a person's friends to be friends with one another.

For a vertex v_i that has at least two neighbors, *clustering coefficient* refers to the fraction of pairs of neighbors of v_i that are themselves neighbors. Thus, the clustering coefficient of a graph quantifies its "cliquishness." The clustering coefficient $C_v(G)$ of a vertex v of graph G is defined by:

$$C_v(G) = \frac{\text{Number of edges between neighbors of } v}{\text{Number of possible edges between neighbors of } v}$$

$$= \frac{|\{(u, w) \in E(G) | (u, v) \in E(G), (w, v) \in E(G)\}|}{\binom{d(v)}{2}},$$

where $E(G)$ is the set of edges of G. Two definitions of the clustering coeffi-cient of the graph are commonly used:

$$C_1(G) = \frac{1}{N} \sum_{v \in V(G)} C_v(G),$$

$$C_2(G) = \frac{3 \times \text{Number of Triangles}}{\text{Number of Connected Triples}}$$

$$= \frac{\sum_{v \in V(G)} \binom{d(v)}{2} C_v(G)}{\sum_{v \in V(G)} \binom{d(v)}{2}},$$

where "connected triples" means three vertices uvw with edges (u, v) and (v, w), the edge (u, w) may or may not be present; and the factor 3 in the numerator accounts for the fact that each triangle is counted three times when the connected triples in the network are counted.

10.2.5 Network Entropy

Entropy is a measure of the uncertainty associated with a random variable. In complex networks, the entropy has been used to characterize properties of the topology, such as the degree distribution of a graph, or the shortest paths between pairs of nodes. As discussed in Chapter 7, for a graph $G = (V, E)$ with a probability distribution P on its vertex set of $V(G)$, where $p_i \in [0, 1]$ is the probability of vertex $v_i, i = 1, \ldots, N$, the entropy is given by:

$$H(G, P) = \sum_{i=1}^{N} p_i \log_2 \left(\frac{1}{p_i} \right) = - \sum_{i=1}^{N} p_i \log_2 (p_i).$$

Entropy is a measure of randomness; the higher the entropy of a network, the more random it is. Thus, a nonrandom network has a zero entropy while a random network has a nonzero entropy. In general, an entropy closer to 0 means more predictability and less uncertainty. For example, consider a Bernoulli random variable X that takes on the values of 1 with probability p and 0 with probability $1-p$. The entropy of the random variable is given by:

$$H(X, p) = -p \log_2(p) - (1-p) \log_2(1-p).$$

Figure 10.1 shows a plot of $H(X, p)$ with p. The function attains its maximum value of 1 at $p = 0.5$, and it is zero at $p = 0$ and at $p = 1$.

10.2.6 Percolation and the Emergence of Giant Component

Percolation theory enables us to assess the robustness of a network. It provides a theoretical framework for understanding the effect of removing nodes in a network. There are many different variants of percolation, but they turn out

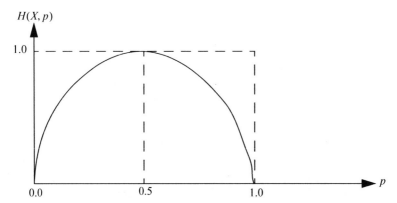

Figure 10.1   Variation of Bernoulli entropy with p.

to be identical in almost all important aspects. As an illustration of a variant of the concept, consider a square grid where each site on the grid is occupied with a probability p. For small values of p we observe mostly isolated occupied sites with occasional pairs of neighboring sites that are both occupied. Such neighboring sites that are both occupied constitute a *cluster*. As p increases, more isolated clusters will form and some clusters grow and some merge. Then at one particular value of p one cluster dominates and becomes very large. This giant cluster is called the *spanning cluster* as it spans the entire lattice. This sudden occurrence of a spanning cluster takes place at a particular value of the occupancy probability known as the *percolation threshold*, p_c, and is a fundamental characteristic of percolation theory. The exact value of the threshold depends on which kind of grid is used and the dimensionality of the grid. Above p_c the other clusters become absorbed into the spanning cluster until at $p = 1$, when every site is occupied.

With respect to networks, percolation addresses the question of what happens when some of the nodes or edges of a network are deleted: What components are left in the graph and how large are they? For example, when studying the vulnerability of the Internet to technical failures or terrorist attacks, one would like to know if there would be a giant component (i.e., disturbances are local) or small components (i.e., the system collapses). Similarly, in the spread of an infectious disease, one would like to limit the spread and would prefer graphs with small components only.

To illustrate the concept of percolation in networks, we start with a graph and independently delete vertices along with all the edges connected to those vertices with probability $1-p$ for some $p \in [0, 1]$. This gives a new random graph model. Assume that the original graph models the contacts where certain infectious disease might have spread. Through vaccination we effectively remove some nodes by making them immune to the disease. We would like to know if the disease still spreads to a large part of the population after the vaccination. Similarly, the failure of a router in a network can be modeled by the removal of the router along with the links connecting the router to other routers in the network. We may want to know the impact of this node failure on the network performance. These are questions that can be answered by percolation theory.

One known result for the emergence of a giant component of a random graph can be stated as follows: The random graph with given vertex degrees $d_1, d_2, \ldots, d_N$, has a giant component if and only if:

$$p > p_c = \frac{\sum_i d_i}{\sum_i d_i(d_i - 1)}.$$

10.3 MODELS OF COMPLEX NETWORKS

In this section we describe different complex network models that have been proposed. These include *random networks*, the *small-world network*, and the *scale-free network*. A random network is a network whose nodes or links or both are created by some random procedure. It is essentially a random graph of the type we discussed earlier. In this network, in the limit that the network size $N \to \infty$, the average degree of each node is given by $E[K] = (N-1)p$. This quantity diverges if p is fixed; thus, p is usually chosen as a function of N to keep $E[K]$ fixed at $p = E[K]/(N-1)$. Under this condition, the limiting degree distribution is a Poisson distribution.

10.3.1 The Small-World Network

The small-world network was introduced by Watts and Strogatz (1998). It is one of the most widely used models in social networks, after the random network. It was inspired by the popular "six degrees of separation" concept, which is based on the notion that everyone in the world is connected to everyone else through a chain of at most six mutual acquaintances. Thus, the world looks "small" when you think of how short a path of friends it takes to get from you to almost everyone else. The small-world network is an interpolation between the random network and the lattice network.

A lattice network is a set of nodes associated with the points in a regular lattice, generally without boundaries. Examples include a one-dimensional ring and a two-dimensional square grid. The network is constructed by adding an edge between every pair of nodes that are within a specified integer lattice radius r. Thus, if $\varphi(\cdot)$ is a measure of lattice distance, the neighbors $\aleph(i)$ of node $i \in V(N)$ are defined by:

$$\aleph(i) = \{j \in V(N) | \varphi(i, j) \le r\},$$

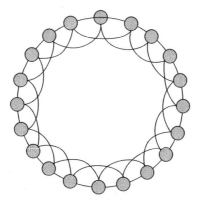

Figure 10.2 One-dimensional regular lattice with a lattice radius of 2. (Reproduced by permission from Macmillan Publishers, Ltd: *Nature*, June 4, 1998.)

where $V(N)$ is the set of nodes of the network. An example of a one-dimensional lattice with $r = 2$ is given in Figure 10.2.

The Watts–Strogatz small-world network is constructed by randomizing a fraction p of the links of the regular network as follows:

- Start with a one-dimensional lattice network with N nodes, such as the one in Figure 10.2.
- *Rewiring Step*: Starting with node 1 and proceeding toward node N, perform the *rewiring procedure* as follows: For node 1, consider the first "forward connection," that is, the connection to node 2. With probability p, delete the link and create a new link between node 1 and some other node chosen uniformly at random among the set of nodes that are not already connected to node 1. Repeat this process for the remaining forward connections of node 1, and then perform this step for the remaining $N-1$ nodes.

If we use $p = 0$ in the rewiring step, the original lattice network is preserved. On the other hand, if we use $p = 1$, we obtain the random network. The small-world behavior is characterized by the fact that the distance between any two vertices is of the order of that for a random network and, at the same time, the concept of neighborhood is preserved, as for regular lattices. In other words, small-world networks have the property that they are locally dense (as regular networks) and have a relatively short path length (as random networks). Therefore, a small-world network is halfway between a random network and a regular network, thus combining properties of local regularity and global disorder. Figure 10.3 illustrates the transition from an ordered (or regular) network to a random network as the rewiring probability p increases.

In the small-world network, both the geodesic distances and the clustering coefficient are functions of the rewiring probability. The geodesic distance

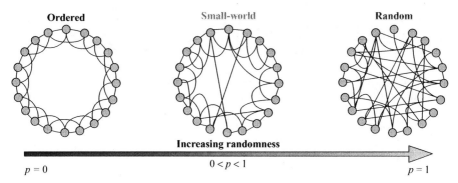

Figure 10.3 Modification of lattice network by increasing rewiring probability.

between nodes increases with the rewiring probability. Similarly, it can be shown that for the Watts–Strogatz small-world network with lattice radius r, the clustering coefficient is given by:

$$C(G) = \frac{3}{2} \left\{ \frac{r-1}{2r-1} \right\}.$$

In summary, a network is said to exhibit small-world characteristics if it has (a) a high amount of clustering and (b) a small characteristic path length. Lattice networks have property (a) but not (b), and random networks almost always have property (b) but not (a).

10.3.2 Scale-Free Networks

The classical random graph has node degrees that are random but with a small probability of having a degree that is much larger than the average. Specifically, the degree distribution is binomial and asymptotically Poisson. Thus, the distribution has exponential tails. Many real-world networks have node degrees that are distributed according to a power law for large degrees; that is, the PMF of the node degree distribution is given by:

$$p_K(k) = Ak^{-\gamma} \quad k = 1, 2, \dots,$$

where γ is some positive constant and A is a normalizing constant. Networks with a power-law degree distribution are called *scale-free networks*.

The mechanism for creating scale-free networks is different from the mechanism for creating random networks and small-world networks. The goal of random networks is to create a graph with correct topological features while the goal of the scale-free network is to capture the network dynamics. Random networks assume that we have a fixed number N of nodes that are then connected or rewired without modifying N. However, most real-world networks evolve by continuous addition of new nodes. That is, real-world networks are not static; they are temporally dynamic. Starting with a small number of nodes, the network grows by increasing the number of nodes throughout the lifetime of the network.

Also, random networks do not assume any dependence of connecting or rewiring probability on the node degree. However, most real-world networks exhibit a *preferential attachment* such that the likelihood of connecting to a node depends on the node's degree. Specifically, when a new node is created, the probability $\pi(k_i)$ that it will be connected to a node v_i with node degree k_i is given by:

$$\pi(k_i) = \frac{k_i}{\sum_j k_j}.$$

Thus, the mechanism used by new nodes in establishing their edges is biased in favor of those nodes that are more highly connected at the time of their arrival. Barabasi and Albert (1999) introduced the mechanisms of growth and preferential attachment that lead to scale-free power-law degree distributions. These features can be summarized as follows. Starting with a small number of nodes, the network evolves by the successive arrival of new nodes that link to some of the existing nodes upon arrival. Assume that we start with the graph G_0 with n_0 nodes and $|E(G_0)|$ edges. Let the initial node set be V_0 and let the node set at step $s > 0$ be V_s. Then,

a. *Growth*: Add a new node v_s to the node set V_{s-1} to generate a new node set: $V_s = V_{s-1} \cup \{v_s\}$

b. *Preferential Attachment*: Add $m \le n_0$ edges to the network at each step, each edge being incident with node v_s and with a node $v_u \in V_{s-1}$ that is chosen with probability:

$$P[\text{Select Node } v_u] = \frac{d(v_u)}{\sum_{v_i \in V_{s-1}} d(v_j)}.$$

c. Stop when n nodes have been added; otherwise repeat the preceding two steps.

It can be shown that after n nodes have been added the probability that vertex v has degree $k \ge m$ is given by:

$$P[k] = \frac{2m(m+1)}{k(k+1)(k+2)}.$$

The average node degree is obtained in van Steen (2010) and given by:

$$\overline{d(G)} = E[k] = \sum_{k=m}^{\infty} kP[k] = 2m(m+1) \sum_{k=m}^{\infty} \frac{k}{k(k+1)(k+2)} = 2m.$$

Also, the clustering coefficient of node v_s after t steps have taken place in the construction of the network, where $s \leq t$, has been derived in Fronczak et al. (2003) as:

$$c_C(v_s) = \frac{m-1}{8(\sqrt{t} + \sqrt{s}/m)^2} \left\{ [\ln(t)]^2 + \frac{4m}{(m-1)^2} [\ln(s)]^2 \right\}.$$

10.4 RANDOM NETWORKS

Recall that we defined a random network as a network whose nodes or edges or both are created by some random procedure. Thus, random networks are modeled by random graphs. Recall that the Bernoulli random graph (or the Erdos–Renyi graph) $G(N, p)$ is a graph that has N vertices such that a vertex is connected to another vertex with probability p, where $0 \leq p \leq 1$. It is assumed that the connections occur independently.

10.4.1 Degree Distribution

If we assume that the edges are independent, the probability that a graph has m edges is $Ap^m (1-p)^{M-m}$, where we assume that self-loops are forbidden, $M = N(N-1)/2$ is the total possible number of edges (as in the complete graph with N vertices, K_n), and $A = \binom{M}{m}$. If we assume that a particular vertex in the graph is connected with equal probability to each of the $N-1$ other vertices and if K denotes the number of vertices that it is connected with, the PMF of K is given by:

$$p_K(k) = P[K = k] = \binom{N-1}{k} p^k (1-p)^{N-1-k} \quad k = 0, 1, \ldots, N-1.$$

Thus, the expected value of K is $E[K] = (N-1)p$, which is the average degree of a vertex. If we hold the mean $(N-1)p = \beta$ constant so that $p = \beta/(N-1)$, then the PMF becomes:

$$p_K(k) = \binom{N-1}{k} \left[\frac{\beta}{N-1} \right]^k \left[1 - \frac{\beta}{N-1} \right]^{N-1-k} \cong \frac{\beta^k}{k!} e^{-\beta} \quad k = 0, 1, \ldots,$$

where the last approximate equality becomes exact in the limit as N becomes large. Since the last quantity is a Poisson distribution, a random graph in which each edge forms independently with equal probability and the degree of each vertex has the Poisson distribution is called a *Poisson random graph*.

10.4.2   Emergence of Giant Component

There is a relationship between the value of p, the size of the largest connected component of the graph, and the *degree distribution* of the graph. (Recall that the degree of a vertex in a graph is the number of edges it has to other vertices. The degree distribution is the probability distribution of these degrees over the whole graph.) Specifically, for smaller values of p, the components are basically of the same size and have a finite mean size. The components follow an exponential degree distribution. At higher values of p, a giant connected component appears in the graph with $O(N)$ vertices. The rest of the vertices in the graph follow an exponential degree distribution. Between these two states, a phase transition occurs at $p = 1/N$. The evolution of the giant component can be illustrated by Figure 10.4 where initially each of the 10 vertices is isolated when $p = 0$. But as p increases, small clusters (or components) are formed. Later on, many of these clusters merge into a giant component.

10.4.3   Connectedness and Diameter

Random networks tend to have small diameters if the connection probability is not too small. This is because as the number of edges increases, there are more paths between node pairs; this in turn provides more opportunity for shorter alternative paths. In Newman (2000), it is shown that the diameter is approximately given by:

$$d = \frac{\ln(N)}{\ln(E[K])} = \frac{\ln(N)}{\ln\{p(n-1)\}},$$

where $E[K]$ is the average node degree.

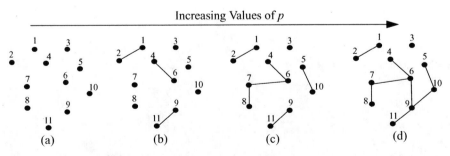

Figure 10.4 Evolution of giant component. (a) $p = p_a = 0$; (b) $0 < p = p_b < 1$; (c) $p_b < p = p_c < 1$; (d) $p_c < p = p_d < 1$.

10.4.4 Clustering Coefficient

Whereas the average path length decreases as p increases because of an increase in the density of edges, the clustering coefficient increases as p increases. This is due to the fact that the clustering coefficient increases with the density because more edges means that more triangular subgraphs are likely to form. In Watts and Strogatz (1998), it is shown that the clustering coefficient of a random network is given by:

$$C(G(N,p)) = \frac{E[K]}{N} = \frac{p(N-1)}{N}.$$

10.4.5 Scale-Free Properties

Recall that the classical random graph has a degree distribution that is binomial and asymptotically Poisson with the result that the distribution has exponential tails. Many real-world networks have node degrees that have a power-law distribution, which is given by:

$$p_K(k) = Ak^{-\gamma} \quad k = 1, 2, \ldots,$$

where γ is some nonnegative constant and A is a normalizing constant. Random networks that have a power-law degree distribution are called *scale-free networks*. Different properties of this class of random networks have been discussed earlier in this chapter.

10.5 RANDOM REGULAR NETWORKS

As defined in Chapter 7, regular graphs are graphs in which each vertex has exactly the same degree. Thus, a regular network is a network in which each node has the same degree. In particular, a k-regular network is one in which the degree of each node is k. Regular networks are highly ordered, as can be seen from the two examples in Figure 10.5.

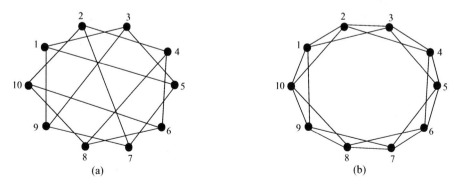

Figure 10.5 Examples of regular networks. (a) $k = 3$; (b) $k = 4$.

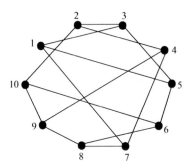

Figure 10.6   Example of a random regular network: $G(10, 3)$.

A random regular graph $G(N, r)$ is a random graph with N vertices in which each vertex has the degree r. Figure 10.6 shows an example of a $G(10, 3)$ graph.

Random regular networks can be used to model social networks in which all individuals have identical numbers of contacts. Observe that the total number of edges is $|E(G)| = Nr/2$. Thus, the network can be constructed only if Nr is even.

There is no natural probabilistic generation procedure of random regular graphs as in the $G(N, p)$ random graph where the generation algorithm creates each possible edge independently with probability p. One of the standard approaches for constructing the graph is the *configuration model* proposed by Bender and Cranfield (1978). This model operates as follows. First, generate n points that constitute the vertices of the graph. To each vertex, attach r stubs, which serve as ends of edges. Next, choose pairs of these stubs uniformly at random from two different nodes and join them to make complete edges. The result is a random regular graph $G(N, r)$.

Random regular networks have been used in Kim and Medard (2004) to model resilience issues in large-scale networks. Specifically, it is shown that if $r \geq 3$, then $G(N, r)$ is r-*connected*, which means that for any pair of vertices i and j, there is a path connecting them in every subgraph obtained by deleting $r - 1$ vertices other than i and j together with their adjacent edges from the graph.

10.6   CONSENSUS OVER RANDOM NETWORKS

Consensus algorithms are used in a wide range of applications, such as load balancing in distributed and parallel computation, and sensor networks. They have also been used as models of opinion dynamics and belief formation in social networks. The central focus of these algorithms is to study whether a group of agents, each of whom has an estimate of an unknown parameter, can reach a global agreement on a common value of the parameter. These agents may engage in information exchange operation such that when agents are made aware of each others' estimates, they may modify their own estimate by taking into account the opinion of other agents. It is assumed that the agents are interconnected by a network in which the agents are the nodes and an edge exists between two nodes i and j if and only if they can exchange information. Thus, by consensus we mean a situation where all agents agree on a common value of the parameter; that is, $x_i = x_j$ for all i and j, where x_i is some parameter available at node i.

With respect to load balancing, the nodes are processors and two nodes i and j are interconnected by an edge if i can send data to j and vice versa. The measurement x_i at node i is the number of jobs the node has to accomplish. To speed up a computation, the nodes exchange information on the x_i along the available edges in order to balance the x_i as much as possible among the nodes.

With respect to sensor networks, the nodes are sensors that are deployed in a geographical area, and they communicate with other sensors in the area over a wireless medium. The quantities x_i that the nodes aim to average can be the measurements taken by each node, which need to be done in order to increase precision by filtering out the noise.

Consider a system of n nodes labeled $1, 2, \ldots, n$ that are interconnected by a network. Each node i at time k holds a value $x_i(k)$. Let $x(k) = [x_1(k), x_2(k), \ldots, x_n(k)]^{\mathsf{T}}$. The interconnection topology of the nodes is represented by a graph in which an edge exists between two nodes if and only if they can exchange information. Let $N(i, k)$ denote the set of nodes that are neighbors of node i at time k. At every time step k, each node i updates its value according to the following equation:

$$x_i(k+1) = \sum_{j \in N(i,k) \cup i} w_{ij}(k) x_j(k),$$

where $w_{ij}(k)$ is the weight that node i assigns to the value of node j, which, as stated earlier, reflects the reliability that node i places in the information from node j. We have that:

$$\sum_{j \in N(i,k) \cup i} w_{ij}(k) = 1 \quad i = 1, \ldots, n.$$

If we define the matrix $W(k)$ whose elements are $w_{ij}(k)$, then the system evolves according to the following equation:

$$x(k+1) = W(k) x(k),$$
$$x(0) = x_0.$$

$W(k)$ is called the *consensus matrix*. A consensus is formed when all the entries of $x(k)$ converge to a common value. That is, the system reaches consensus in probability if, for any initial state $x(0)$ and any $\varepsilon > 0$,

$$P[|x_i(k) - x_j(k)| > \varepsilon] \to 0,$$

as $k \to \infty$ for all $i, j = 1, \ldots, n$. This is a weak form of reaching a consensus. A stronger form is reaching a consensus almost surely, which is that for any initial state $x(0)$,

$$|x_i(k) - x_j(k)| \to 0,$$

as $k \to \infty$ for all $i, j = 1, \ldots, n$.

To further motivate the discussion, we first consider consensus over a fixed network and then discuss consensus over a random network.

10.6.1   Consensus over Fixed Networks

One of the simplest models of consensus over fixed networks is that proposed by DeGroot (1974). In this model, each node i starts with an initial probability $p_i(0)$, which can be regarded as the probability that a given statement is true. Let $p(0) = [p_1(0), p_2(0), \ldots, p_n(0)]^T$. The $n \times n$ consensus matrix $W(k)$ has elements $w_{ij}(k)$ that represent the weight or trust that node i places on the current belief of node j in forming node i's belief for the next period. In particular, $W(k)$ is a stochastic matrix that does not depend on k so that we may write $W(k) = W$. Beliefs are updated over time as follows:

$$p(k+1) = Wp(k).$$

Solving this equation recursively we obtain:

$$p(k) = W^k p(0).$$

Thus, the algorithm is as follows: Every node runs a first-order linear dynamical system to update its estimation using the consensus matrix W and the initial probability vector $p(0)$. The updating process converges to a well-defined limit if $\lim_k W^k p(0)$ exists for all $p(0)$, which is true if W is an irreducible and aperiodic matrix. Thus, any closed group of nodes reaches a consensus if the consensus matrix W is irreducible and aperiodic.

Example 10.1: Consider the three-node network shown in Figure 10.7 where the transition probabilities are the w_{ij}. The network can be interpreted as follows. Agent 1 weighs all estimates equally; agent 2 weighs its own estimate and that of agent 1 equally and ignores that of agent 3; and agent 3 weighs its own estimate equally with the combined estimates of agents 1 and 2.

Since the consensus matrix,

$$W = \begin{bmatrix} 1/3 & 1/3 & 1/3 \\ 1/2 & 1/2 & 0 \\ 1/6 & 1/3 & 1/2 \end{bmatrix},$$

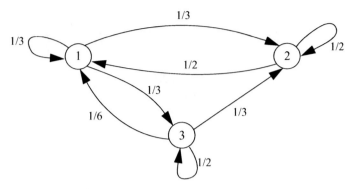

Figure 10.7 Figure for Example 10.1.

is an irreducible and aperiodic Markov chain, the limits $\pi_j = \lim_{k \to \infty} w_{ij}(k)$ exist
and are independent of the initial distribution, where:

$$\pi = \pi W,$$
$$1 = \pi 1,$$
$$\pi = [\pi_1, \pi_2, \ldots, \pi_n],$$
$$1 = [1, 1, \ldots 1]^T.$$

Assume that $p(0) = [1, 0, 0]^T$. Thus, we have that:

$$\pi_1 = \frac{1}{3}\pi_1 + \frac{1}{2}\pi_2 + \frac{1}{6}\pi_3,$$
$$\pi_2 = \frac{1}{3}\pi_1 + \frac{1}{2}\pi_2 + \frac{1}{3}\pi_3,$$
$$1 = \pi_1 + \pi_2 + \pi_3.$$

The solution to this system of equations is:

$$\pi = [\pi_1, \pi_2, \pi_3] = [9/25, 10/25, 6/25].$$

Thus,

$$\lim_{k \to \infty} W^k = \begin{bmatrix} \pi_1 & \pi_2 & \pi_3 \\ \pi_1 & \pi_2 & \pi_3 \\ \pi_1 & \pi_2 & \pi_3 \end{bmatrix} = \begin{bmatrix} 9/25 & 10/25 & 6/25 \\ 9/25 & 10/25 & 6/25 \\ 9/25 & 10/25 & 6/25 \end{bmatrix},$$

$$\lim_{k \to \infty} p(k) = \begin{bmatrix} 9/25 & 10/25 & 6/25 \\ 9/25 & 10/25 & 6/25 \\ 9/25 & 10/25 & 6/25 \end{bmatrix} \begin{bmatrix} 1 \\ 0 \\ 0 \end{bmatrix} = \begin{bmatrix} 9/25 \\ 9/25 \\ 9/25 \end{bmatrix}.$$

That is, all three nodes will converge on the same value of $p_i(k)$ as $k \to \infty$, which
is 9/25.

10.6.1.1   Time to Convergence in a Fixed Network   We have discussed the condition for the convergence of a consensus algorithm in a fixed network. However, sometimes it is important to know the time to convergence of an algorithm. To do this we need to know how the opinions at time k, $p(k) = W^k p(0)$, differ from the limiting opinions, $p(\infty) = W^\infty p(0)$. One of the ways we can keep track of this difference is by tracking the difference between W^k and W^∞. To do this, we use the *diagonalization method* that gives:

$$W = S\Lambda S^{-1},$$
$$W^k = S\Lambda^k S^{-1},$$

where S is the matrix whose columns are the eigenvectors of W, and Λ is the diagonal matrix of the corresponding eigenvalues of W. The eigenvalues λ of W satisfy the equation:

$$|\lambda I - W| = 0,$$

where $\lambda = [\lambda_1, \lambda_2, \dots, \lambda_n]$. Let X_i be the eigenvector associated with λ_i. Then we have that:

$$\Lambda = \begin{bmatrix} \lambda_1 & 0 & 0 & \dots & 0 \\ 0 & \lambda_2 & 0 & \dots & 0 \\ 0 & 0 & \dots & \dots & 0 \\ \dots & \dots & \dots & \dots & \dots \\ 0 & 0 & 0 & \dots & \lambda n \end{bmatrix},$$

and

$$(\lambda_i I - W) X_i = 0 \quad i = 1, \dots, n,$$
$$X_i = [x_{1i}, x_{2i}, \dots, x_{ni}]^T,$$
$$S = [X_1, X_2, \dots, X_n].$$

The Perron–Frobenius theorem states that for an aperiodic and irreducible stochastic matrix P, 1 is a simple eigenvalue and all other eigenvalues have absolute values that are less than 1. Moreover, the left eigenvector X_1 corresponding to eigenvalue 1 can be chosen to have all the positive entries and thus can be chosen to be a probability vector by multiplying with an appropriate constant. Thus, for the equation $p(k) = W^k p(0)$, we have that:

$$[W^k]_{ij} = [X_1]_j + \sum_{l=2}^{n} \lambda_l^k [S]_{il} [S^{-1}]_{lj}.$$

From this we obtain:

$$p_i(0) - p_i(\infty) = \sum_j p_j(0) \sum_{l=2}^{n} \lambda_l^k [S]_{il} [S^{-1}]_{lj}.$$

This means that the convergence of W^k to W^∞ depends on how quickly λ_2^k goes to zero, since $|\lambda_2|$ is the second largest eigenvalue.

10.6.2 Consensus over Random Networks

Consensus over fixed networks assumes that the network topology is fixed. A variant of the problem is when either the topology of the network or the parameters of the consensus algorithm can randomly change over time. In this section we consider the case where the network topology changes over time. This model has been analyzed in Hatano and Meshahi (2005) and in Tahbaz-Salehi and Jadbabaie (2008).

To deal with the case where the network topology changes over time, we consider the set of possible graphs that can be generated among the n nodes. Each such graph will be referred to as an *interaction graph*. Assume that there is a set of interaction graphs, which are labeled $G_1, G_2, \ldots,$ and let $G = \{G_1, G_2, \ldots\}$ denote the set of interaction graphs. We assume that the choice of graph G_i at any time step k is made independently with probability p_i. Let the consensus matrix corresponding to graph G_i be W_i, and let the set of consensus matrices be $W = \{W_1, W_2, \ldots\}$. Thus, the state $x(k)$ now evolves stochastically, and the convergence of the nodes to the average consensus value will occur in a probabilistic manner. We say that the system reaches a consensus asymptotically on some path $\{W(k)\}_{k=1}^{\infty}$ if along the path there exists $x^* \in R$ such that $x_i(k) \to x^*$ for all i as $k \to \infty$. The value x^* is called the asymptotic consensus value.

Let $w_{ij}^{(m)}(k)$ be the weight that node i assigns to the value of node j under the interaction graph G_m in time step k, where $m = 1, 2, \ldots$. Assume that the interaction graphs are used in the order $G_1, G_2, \ldots, G_k$ in the first k time steps. Then the system evolution,

$$x(k+1) = W(k)x(k),$$
$$x(0) = x_0,$$

becomes:

$$x(k) = W_k W_{k-1} \ldots W_1 x(0) = \Psi(k)x(0).$$

Thus, in order to check for an asymptotic consensus, we need to investigate the behavior of infinite products of stochastic matrices. Suppose that the average consensus matrix $E[W(k)]$ has n eigenvalues that satisfy the condition:

$$0 \geq |\lambda_n(E[W(k)]])| \leq \lambda_{n-1}(E[W(k)]) \leq \ldots \leq |\lambda_2(E[W(k)]])| \leq |\lambda_1(E[W(k)]])| = 1.$$

In Tahbaz-Salehi and Jadbabaie (2008), it is shown that the discrete dynamical system $x(k + 1) = W(k)x(k)$ reaches state consensus almost surely if $|\lambda_2(E[W(k)]])| < 1$.

10.7 SUMMARY

Complex systems have traditionally been modeled by random graphs that do not produce the topological and structural properties of real networks. Many new network models have been introduced in recent years to capture the scale-free structure of real networks such as the WWW, Internet, biological networks, and social networks. Several properties of these networks have been discussed including average path length, clustering coefficient, and degree distribution. While some complex systems, such as the power grid, can still be modeled by the exponential degree distribution associated with the traditional random graph, others are better modeled by a power-law distribution; sometimes a mixture of exponential and power-law distributions may be used.

The study of complex networks has given rise to a new network science that applies graph theory to the study of dynamic systems that occur in diverse fields. It is an evolving field and the goal of this chapter has been to present its basic principles.

10.8 PROBLEMS

10.1 Consider the graph shown in Figure 10.8.
 a. Calculate the diameter of the graph.
 b. Obtain the closeness centrality of vertex 1.
 c. Obtain the betweenness centrality of vertex 1.
 d. Obtain the clustering coefficient of vertex 1.

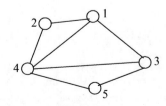

Figure 10.8   Figure for Problem 10.1.

10.2 Let $d(v_i)$ denote the degree of vertex v_i of a graph, and let p_i be the probability distribution on its vertex set, where:

$$p_i = \frac{d(v_i)}{\sum_k d(v_k)}.$$

Using this definition of probability distribution, obtain the entropy of the graph in Figure 10.8.

10.3 Consider a Watts–Strogatz small-world network with lattice radius 3. Calculate the clustering coefficient of the network.

10.4 Calculate the clustering coefficient and the approximate diameter of the random network $G(10, 0.3)$.

10.5 Use the configuration model to generate three instances of the random regular network $G(8, 4)$.

REFERENCES

Albert, R. and A.-L. Barabasi (2002). "Statistical Mechanics of Complex Networks," *Reviews of Modern Physics*, vol. 74, pp. 47–97.

Appel, K. and W. Haken (1976). "Every Planar Map is Four Colorable," *Bulletin of the American Mathematical Society*, vol. 82, pp. 711–712.

Appel, K. and W. Haken (1977). "The Solution of the Four-Color-Map Problem," *Scientific American*, vol. 237, pp. 108–121.

Barabasi, A.-L. and R. Albert (1999). "Emergence of Scaling in Random Networks," *Science*, vol. 286, pp. 509–512.

Barrat, A., M. Bathelemy and A. Vespignani (2008). *Dynamical Processes on Complex Networks*, Cambridge University Press, Cambridge, UK.

Baskett, F., K.M. Chandy, R.R. Muntz and F.G. Palacios (1975). "Open, Closed and Mixed Networks of Queues with Different Classes of Customers," *Journal of the ACM*, vol. 22, pp. 248–260.

Bender, E.A. and E.R. Cranfield (1978). "The Asymptotic Number of Labeled Graphs with Given Degree Sequences," *Journal of Combinatorial Theory, Series A*, vol. 24, pp. 296–307.

Besag, J. (1974). "Spatial Interaction and the Statistical Analysis of Lattice Systems," *Journal of the Royal Statistical Society, Series B*, vol. 36, pp. 192–236.

Boccaletti, S., V. Latora, Y. Moreno, M. Chavez and D.-U. Hwang (2006). "Complex Networks: Structure and Dynamics," *Physics Reports*, vol. 424, pp. 175–308.

Burke, P.J. (1956). "The Output of a Queueing System," *Operations Research*, vol. 4, pp. 699–704.

Fundamentals of Stochastic Networks, First Edition. Oliver C. Ibe.
© 2011 John Wiley & Sons, Inc. Published 2011 by John Wiley & Sons, Inc.

Buzen, J.P. (1973). "Computational Algorithms for Closed Queueing Networks with Exponential Servers," *Communications of the ACM*, vol. 16, pp. 527–531.

Chao, X., M. Miyazawa and M. Pinedo (1999). *Queueing Networks: Customers, Signals and Product Form Solutions*, John Wiley, Chichester, UK.

Costa L. Da F., F.A. Rodrigues, G. Travieso and P.R. Villas Boas (2007). "Characterization of Complex Networks: A Survey of Measurements," *Advances in Physics*, vol. 56, pp. 167–242.

DeGroot, M.H. (1974). "Reaching a Consensus," *Journal of the American Statistical Association*, vol. 69, pp. 118–121.

Dorogovtsev, S.N. and J.F.F. Mendes (2002). "Evolution of Networks," *Advances in Physics*, vol. 51, pp. 1079–1187.

Dorogovtsev, S.N., A.V. Goltsev and J.F.F. Mendes (2008). "Critical Phenomena in Complex Networks," *Reviews of Modern Physics*, vol. 80, pp. 1275–1335.

Doshi, B.T. (1986). "Queueing Systems with Vacations—A Survey," *Queueing Systems*, vol. 1, pp. 29–66.

Durrett, R. (2007). *Random Graph Dynamics*, Cambridge University Press, Cambridge, UK.

Easley, D. and J. Kleinberg (2010). *Networks Crowds and Markets: Reasoning about a Highly Connected World*, Cambridge University Press, Cambridge, UK.

Erdos, P. and A. Renyi (1959). "On Random Graphs," *Publication of the Mathematical Institute of the Hungarian Academy of Sciences*, vol. 5, pp. 17–61.

Erdos, P. and A. Renyi (1960). "On the Evolution of Random Graphs," *Publicationes Mathematicae Debrecen*, vol. 6, pp. 209–297.

Forney, G.D. (2001). "Codes on Graphs: Normal Realizations," *IEEE Transactions on Information Theory*, vol. 47, pp. 520–548.

Frank, O. and D. Strauss (1986). "Markov Graphs," *Journal of the American Statistical Association*, vol. 47, pp. 832–842.

Fronczak, A., P. Fronczak and J. Holyst (2003). "Mean-field Theory for Clustering Coefficients in Barabasi-Albert Networks," *Physical Review E, Statistical, Nonlinear, and Soft Matter Physics*, vol. 68, p. 046126.

Gaver, D.P. (1968). "Diffusion Approximations and Models for Certain Congestion Problems," *Journal of Applied Probability*, vol. 5, pp. 607–623.

Gaver, D.P. (1971). "Analysis of Remote Terminal Backlogs under Heavy Demand Conditions," *Journal of the ACM*, vol. 18, pp. 405–415.

Gelenbe, E. and G. Pujolle (1998). *Introduction to Queueing Networks*, 2nd ed., John Wiley, Chichester, UK.

Gelenbe, E., P. Glynn and K. Sigman (1991). "Queues with Negative Arrivals," *Journal of Applied Probability*, vol. 28, pp. 245–250.

Glynn, P.W. (1990). "Diffusion Approximations," Chapter 4 in *Stochastic Models*, D.P. Heyman and M.J. Sobel, editors, North-Holland, Amsterdam.

Gordon, W.J. and G.F. Newell (1967). "Closed Queueing Systems with Exponential Servers," *Operations Research*, vol. 15, pp. 254–265.

Harrison, J.M. (1985). *Brownian Motion and Stochastic Flow Systems*, John Wiley, New York.

Hatano, Y. and M. Meshahi (2005). "Agreement over Random Networks," *IEEE Transactions on Automatic Control*, vol. 50, pp. 1867–1872.

Heyman, D.P. (1975). "A Diffusion Model Approximation for the GI/G/1 Queue in Heavy Traffic," *Bell System Technical Journal*, vol. 54, pp. 1637–1646.

Huang, C. and A. Darwiche (1996). "Inference in Belief Networks: A Procedural Guide," *International Journal of Approximate Reasoning*, vol. 15, pp. 225–263.

Ibe, O.C. (2005). *Fundamentals of Applied Probability and Random Processes*, Elsevier Academic Press, Burlington, MA.

Ibe, O.C. (2009). *Markov Processes for Stochastic Modeling*, Elsevier Academic Press, Burlington, MA.

Ibe, O.C. and J. Keilson (1995). "Multiserver Threshold Queues with Hysteresis," *Performance Evaluation*, vol. 21, pp. 185–213.

Ibe, O.C. and K.S. Trivedi (1990). "Stochastic Petri Net Models of Polling Systems," *IEEE Journal on Selected Areas in Communications*, vol. 8, pp. 1646–1657.

Ibe, O.C. and K.S. Trivedi (1991). "Stochastic Petri Net Analysis of Finite-Population Vacation Queueing Systems," *Queueing Systems*, vol. 8, pp. 111–128.

Ibe, O.C., K.S. Trivedi and H. Choi (1993). "Performance Evaluation of Client-Server Systems," *IEEE Transactions on Parallel and Distributed Systems*, vol. 4, pp. 1217–1229.

Jackson, M.O. (2008). *Social and Economic Networks*, Princeton University Press, Princeton, NJ.

Kauffman, S.A. (1969). "Metabolic Stability and Epigenesis in Randomly Constructed Genetic Nets," *Journal of Theoretical Biology*, vol. 22, pp. 437-467.

Keilson, J., J. Cozzolino and H. Young (1968). "A Service System with Unfilled Requests Repeated," *Operations Research*, vol. 16, pp. 1126–1137.

Kim, M. and M. Medard (2004). "Robustness in Large-Scale Random Networks," *Proceedings of the IEEE INFOCOM 2004*, vol. 4, pp. 2364–2373.

Kleinrock, L. (1975). *Queueing Systems Volume 1: Theory*, John Wiley, New York.

Kobayashi, H. (1974). "Application of the Diffusion Approximation to Queueing Networks I: Equilibrium Queue Distributions," *Journal of the ACM*, vol. 21, pp. 316–328.

Lewis, T.G. (2009). *Network Science: Theory and Applications*, John Wiley, Hoboken, NJ.

Little, J.D.C. (1961). "A Proof for the Queueing Formula L = λW," *Operations Research*, vol. 9, pp. 383–387.

Marchal, W.G. (1978). "Some Simpler Bounds on the Mean Queueing Time," *Operations Research*, vol. 26, pp. 1083–1088.

Marshall, K.T. (1968). "Some Inequalities on Queueing," *Operations Research*, vol. 16, pp. 651–665.

Newell, G.F. (1971). *Applications of Queueing Theory*, Chapman and Hall, London.

Newman, M.E.J. (2000). "Models of the Small World," *Journal of Statistical Physics*, vol. 101, pp. 819–841.

Newman, M.E.J. (2002). "The Structure and Function of Complex Networks," *SIAM Review*, vol. 45, pp. 167–256.

Newman, M.E.J. (2010). *Networks: An Introduction*, Oxford University Press, Oxford, UK.

Papadimitriou, C.H. and K. Steiglitz (1982). *Combinatorial Optimization: Algorithms and Complexity*, Prentice-Hall, Englewood Cliffs, NJ.

Posner, M. and B. Bernholtz (1968). "Closed Finite Queueing Networks with Time Lags and with Several Classes of Units," *Operations Research*, vol. 16, pp. 977–985.

Reich, E. (1957). "Waiting Times When Queues Are in Tandem," *Annals of Mathematical Statistics*, vol. 28, pp. 768–773.

Reiser, M. and S.S. Lavenberg (1980). "Mean Value Analysis of Closed Multichain Queueing Networks," *Journal of the ACM*, vol. 27, pp. 313–322.

Saaty, T.L. (1972). "Thirteen Colorful Variations on Guthrie's Four-Color Conjecture," *American Mathematical Monthly*, vol. 79, pp. 2–43.

Tahbaz-Salehi, A. and A. Jadbabaie (2008). "A Necessary and Sufficient Condition for Consensus over Random Networks," *IEEE Transactions on Automatic Control*, vol. 53, pp. 791–795.

Takagi, H. (1991). *Queueing Analysis—A Foundation of Performance Analysis Volume 1: Vacation and Priority Systems*, North-Holland, Amsterdam, the Netherlands.

Van Steen, M. (2010). *Graph Theory and Complex Networks: An Introduction*, Maarten van Steen, Amsterdam, the Netherlands.

Vega-Redondo, F. (2007). *Complex Social Networks*, Cambridge University Press, Cambridge, UK.

Watts, D.J. and S.H. Strogatz (1998). "Collective Dynamics of 'Small-World' Networks," *Nature*, vol. 393, pp. 440–442.

Wiberg, N. (1996). *Codes and Decoding on General Graphs*, Linkoping Studies in Science and Technology, Dissertation 440, Linkoping University, Linkoping, Sweden.

Wolff, R.W. (1982). "Poisson Arrivals See Time Averages," *Operations Research*, vol. 30, pp. 223–231.

Wolff, R.W. (1989). *Stochastic Modeling and the Theory of Queues*, Prentice Hall, Englewood Cliff, NJ.

INDEX

Fundamentals of Stochastic Networks, First Edition. Oliver C. Ibe.
© 2011 John Wiley & Sons, Inc. Published 2011 by John Wiley & Sons, Inc.